The Bewitchment of Silver

This series of publications on Africa, Latin America, and Southeast Asia is designed to present significant research, translation, and opinion to area specialists and to a wide community of persons interested in world affairs. The editor seeks manuscripts of quality on any subject and can generally make a decision regarding publication within three months of receipt of the original work. Production methods generally permit a work to appear within one year of acceptance. The editor works closely with authors to produce a high quality book. The series appears in a paperback format and is distributed worldwide. For more information, contact the executive editor at Ohio University Press, Scott Quadrangle, University Terrace, Athens, Ohio 45701.

Executive editor: Gillian Berchowitz
AREA CONSULTANTS
Africa: Diane Ciekawy
Latin America: Thomas Walker
Southeast Asia: William H. Frederick

The Monographs in International Studies series is published for the Center for International Studies by the Ohio University Press. The views expressed in individual volumes are those of the authors and should not be considered to represent the policies or beliefs of the Center for International Studies, the Ohio University Press, or Ohio University.

The Bewitchment of Silver

The Social Economy of Mining in Nineteenth-Century Peru

José R. Deustua

Ohio University Center for International Studies
Monographs in International Studies
Latin America Series No. 31
Athens

The books in the Center for International Studies Monograph Series
are printed on acid-free paper ∞

08 07 06 05 04 03 02 01 00   5 4 3 2 1

Cover art: Mining guano workers in late-nineteenth-century Peru.
Pictures and illustrations are taken from Carlos B. Cisneros, *Atlas del Perú: Político,*
Minero, Agrícola, Industrial y Comercial (Lima: Librería e Imprenta Gil, 1902).

Library of Congress Cataloging-in-Publication

Deustua, José.
   The bewitchment of silver : the social economy of mining in nineteenth-century
Peru / José R. Deustua.
      p.   cm. — (Monographs in international studies. Latin America series ;
ISSN no. 31)
   Includes bibliographical references and index.
   ISBN 0-89680-209-4 (pbk. : alk. paper)
   1. Silver mines and mining—Peru—History—19th century.   2. Silver min-
ers—Peru—History—19th century.   3. Peru—Economic conditions—19th cen-
tury.   I. Title.   II. Title: Social economy of mining in nineteenth-century
Peru.   III. Title: Social economy of mining in 19th century Peru.   IV. Series.
HD9537.P42D48   1999
338.2'7421'098509034—dc21   99-34816

To Sam,
flor de invernadero en pleno verano.
To Inés,
amapolas de otoño.
To José,
jacintos salvajes.

Pimancha huillaiman
cay huaccha causita?
Mamaiman huillaiman
huajapalla huaman.

—folk song of peasant miners from the central sierra of Peru,
sung in Wanka Quechua

Auri sacra fames!

—Virgil, *Aeneid*

To whom can I tell
all this orphan sadness?
If I tell my mother
she will start to cry.

Execrable thirst for gold!

Contents

Illustrations

Maps

Figures

Photographs
Following page 99

Tables

Acknowledgments

This book is one of the main results of a project that started almost two decades ago. My memories are still attached to a late spring day in Lima, Peru, when Heraclio Bonilla invited us for a tasty lunch at a Spanish restaurant in Jorge Washington's square, close to the Avenida Arequipa. We were Carlos Sempat Assadourian, Tristan Platt, Carlos Contreras, I, and, of course, Heraclio. A new research project was being elaborated at the Instituto de Estudios Peruanos whose topic was mining in the Andean countries from the sixteenth to the twentieth centuries. That lunch took place eighteen years ago. Since then, many books and articles have been published by the *comensales*. This new book is also a result of that collective commitment to knowledge, although my association with Heraclio's project ended in 1983. At that time I moved to Paris, where I continued my doctoral studies. In France I worked at the Ecole des Hautes Etudes en Sciences Sociales under the direction of Professor Ruggiero Romano. Ruggiero was my Directeur d'Etudes and I greatly appreciate his help, which made me a better historian. I would also like to thank Professor Nathan Wachtel, a member of my thesis committee and frequent advisor, from whom I learned that the Andes are part of a worldwide *anthropologie historique*. I thank, in the United States, the departments of History at Florida International University (FIU), the University of Miami, Stanford University, and the University of Washington, where I was a visiting professor from Latin America. With their conversation, friendship, and their own careers, Mark Szchuman at FIU, Steve Stein at Miami, Steve Haber at Stanford, and Charles Bergquist, Dauril Alden, and Carlos Gil at the University of Washington were the first to give me an insight into what it means to be a historian in the United States. These were valuable lessons for a tenured professor, a *profesor auxiliar* at the Universidad de San Marcos in Lima. I also thank the Latin American Studies Program and the Department of History at the University of Illinois at Chicago (UIC), where I teach in these days. Particularly I thank my friends and colleagues there—Bruce Calder, Marc Zimmerman, Mary Kay Vaughan, Leo Schelbert, Marion S. Miller, Rafael Núñez Cedeño, David Badillo, Víctor Ortiz, and Otto Pikaza—for their example, advice, and encouragement. I also received from UIC a three-year research minority grant that allowed me to travel several times to Peru to finish the book. Susan Stokes helped me with the English composition of previous versions of this book. While writing and continuing to perform my household duties I enjoyed immensely the company of my son, Samuel Daniel Deustua S. Last but not least I must acknowledge the hard work and probing questions of my graduate and undergraduate students at UIC and other centers of learning, especially Pamela Baker, Ana María Kapeluzs Poppi, and Benson Stein. Their questions and comments during the usual presentation of my research material and historical ideas, particularly in my seminar Historical Capitalism in Latin America and Europe, were a constant reminder that a book is a living thing. Made up of questions, work, and experiences, it should generate more questions, more work, and more experiences.

The Bewitchment of Silver

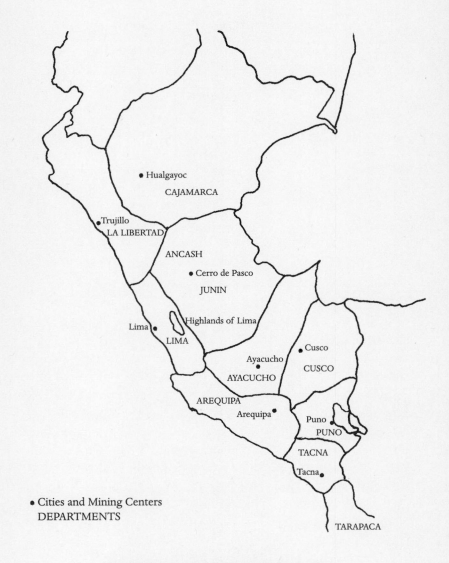

• Hualgayoc
CAJAMARCA

Trujillo
LA LIBERTAD

ANCASH

• Cerro de Pasco
JUNIN

Highlands of Lima
Lima •
LIMA

• Cusco
Ayacucho
CUSCO
AYACUCHO

AREQUIPA
Arequipa •

Puno •
PUNO

TACNA

Tacna •

• Cities and Mining Centers
DEPARTMENTS

TARAPACA

Main Mining Regions in Nineteenth-Century Peru

Chapter 1

Introduction

THIS BOOK IS an inquiry into the impact of an export sector on the evolving national economy in a country at the periphery of developing world capitalism in the nineteenth century. The export sector is mining, the country Peru. I have focused on mining because it was crucial for nineteenth-century Peru, although badly understood by current historiography. I argue here that mining was not only a key export sector for the economies of Peru and the world, but that its role was decisive in the creation of an internal market and for the development of Peru as a nation. My most general claim in this book is that export sectors in peripheral countries must be grasped not only in terms of their role in generating foreign exchange and in linking a country with the world economy, but also in terms of the difficult task of achieving domestic development. Thus, this book is deeply concerned with the question of market building and domestic development in nineteenth-century Latin America.

This study aims to be empirical. I have tried to assess thoroughly the real importance of mining for the society and national economy of nineteenth-century Peru. But I also found it necessary to address a previous and broader question. There is a current debate in the social sciences about the role of knowledge and the reconstruction of facts and social situations for historical analysis. The influence of postmodernism is increasing, focusing particularly on political and cultural aspects of historical reality.[1] Marxist, positivist, and economic analysis in general, whether Keynesian or neoliberal, seems to have lost ground. This book also tries to engage in the debate that postmodernism and poststructuralism have opened concerning larger theoretical questions in the social sciences. However, I am not trying to elaborate a total conceptual theory here. I only present some preliminary thoughts that relate to the present state of the historical discipline in the United States.[2] I also introduce a new concept for the social and historical sciences in today's world, that of social economy. Economics at the beginning of the nineteenth century was a philosophical discipline, linked even to morals and ethics.[3] By the end of the century, however, economics was becoming the abstract, highly mathematical discipline it is today. Economics no longer relates to the social sciences; it is merely a part of business administration. The idea of a social economy, therefore, supposes a break with this contemporary tradition and tries to integrate economics with the twentieth-century achievements of the other social and historical sciences to bring this discipline back to its classical roots—to Adam Smith and Karl Marx, but also to John Maynard Keynes and Simon Kuznets.

Thus, throughout the book the reader will notice my questioning of the validity of economic and quantitative analysis, of the role of social history in understanding historical events, and of the importance of the individual and of social classes as actors in the human drama. I do not pretend, of course, to have all the answers to these deep and complex questions. This is just a first step in the elaboration of such a theory.

Historical Questions

Aristotle began his *Metaphysics* by stating that all people "naturally have an impulse to get knowledge."[4] Antonio Gramsci, the southern Italian socialist intellectual and politician who spent most of his productive years in Mussolini's jails, wrote that all men (and women) are philosophers.[5] Both authors express the same idea, the importance of attaining knowledge, of being philosophers, of having an understanding of the world that surrounds us. This book is an attempt to get knowledge, to reflect, to think about the meaning of history. However, it is not a book about philosophy. It is rather a history book, a book about social and economic history.

This matter raises the problem of the specialization of disciplines (philosophy, history, social and economic history) that could also be stated in terms of the specialization of geographic areas of study (Latin America versus Europe, the United States versus Argentina or Brazil, etc.). As our knowledge becomes increasingly more precise, more specific to our own area of study, our own subdiscipline, we lose global perceptions, universal aspirations. Although this is a book about a specific society at a specific time, viewed through a specific discipline (social and economic history), I have tried not to make it a specialized book. The questions I ask aim to be universal. While writing about the Peruvian silver mines I was wondering, at the same time, about U.S. mine workers in West Virginia, or copper mines in Arizona, or the huge coal mines, almost citadels, in Westphalia or in the Pas de Calais in France. In this sense, too, history touches philosophy and Gramsci encounters Aristotle, just as Garcilaso de la Vega, with his notes on the mines and the mineral works at the end of Inca times, encounters Mariano de Rivero.[6] Every time general questions are asked, historical inquiry follows philosophical lines. In other words, the writing of history always presupposes a theoretical framework, despite the goal of objectivity one assumes.[7] I thus started with a very large question: What was the contribution of mining to any particular society in any specific period of time? But I had to concentrate my forces and focus on just one society (Peru), in a specific period of time (the nineteenth century).

Mining has been a crucial economic activity since the earliest human history. According to some archeological schools the use of mining resources and their transformation into human tools divided the evolution of human societies from "the age of copper to the ages of bronze and iron."[8] However, modern mining has experienced two dramatic transitions. The first was the European discovery and conquest of America and the creation of a true world economy. The Spanish and Portuguese exhausted Latin American mines, shifting large quantities of gold and silver to Europe, and setting off, in the words of Earl Hamilton, a "price revolution"—and, I would add, a step further toward capitalism.[9] Mexico, Colombia, Brazil, and the viceroyalty of Peru, which for a few centuries included the present-day countries of Bolivia and Chile, were the principal sources of this fantastic new wealth directed toward the European economies.

The second watershed transition was the Industrial Revolution of the late eighteenth and early nineteenth centuries. Here, instead of a territorial expansion, we have a technical and economic revolution.[10] How and to what extent did Latin America become enmeshed in the web of the British Industrial Revolution? In the web of European capitalism?

Many answers have been given in the past to this critical problem.[11] A debate has ensued about whether relations between Britain and Latin America constituted formal or informal imperialism in the nineteenth century, and about the magnitude of British influence in the region. D. C. M. Platt argues that foreign influence was only modest in Latin American economic development and, therefore, the impact of the British Industrial Revolution was limited. Stanley Stein counters that the influence of British and of foreign interests in general was powerful, creating severe dependency relations.[12] The data presented in this book contribute to this debate, although my study focuses rather on the domestic impact of the mining export economy; this, then, is an internal account of an internationally connected export sector.

The Economic Development of Mining in Peru and Latin America

This same discussion has had repercussions in the historiography of nineteenth-century Peru. Here, because one perspective had emphasized the powerful influence of international markets and capital in the failure of the guano boom of the 1840s to the 1870s to generate sustained development, this perspective has been called *dependentista* by those who argue that protectionism and nationalism had their social origins in early nineteenth-century Peru.[13] Similarly, one recent book underscores the early though unsuccessful influences of British, French, and U.S. consular and trade agents in Peru to shape commercial policies (unsuccessful because of the rather larger power of protectionist and nationalist forces). But another study (of peasants in Peru's central highlands) situates the real penetration of foreign capital in the late nineteenth century.[14]

In the Platt-Stein debate, however, both sides emphasized the condition of Latin America mostly as a market for British goods, particularly from the textile industry. From that starting point the discussion has centered on the role of foreign merchants in that trade.[15] This book focuses more on the specific aspects of production, especially mining production. Any study of the productive system should reflect more closely the real economy, whereas trade, by definition, is related to exchange, to the exchange of goods within an economy or, rather, as in the Platt-Stein debate, between a national and a foreign economy. Trade is part, then, of a different dimension of the economic cycle, one that belongs rather to the commercial or financial aspects of the economy. My interest, on the contrary, will be centered on production.

Thus, narrowing still more the scope of my inquiry, this book will try to answer the following questions: What was the structure of the mining sector in nineteenth-century Peru? What was the contribution of mining to the national economy and society? What was its quantitative and qualitative importance? What were the real dimensions of this specific economic sector within the nineteenth-century Peruvian economy?

Was mining an export economy throughout the century? Did it become more so during any specific period of time? During a specific sectoral transition? From silver to copper mining, for example? How did mining promote domestic development? What relationships did it have with the constitution of an internal market in nineteenth-century Peru? How was it linked with other economic sectors? And, finally, how did trading and transportation shape the mining sector for domestic development or as an export economy, or both?[16]

I will demonstrate that, in addition to its role as a critical export sector, mining was a constant engine of development of the Peruvian national economy. It created a dynamic, although narrow, internal market, through the use of the muleteering system. These conditions, however, changed with the introduction of the railroad.

The main goal of this book is, therefore, to prove the consistency of the Peruvian mining industry throughout the nineteenth century. Of course mining had moments of growth, such as the 1830s and 1840s in the silver sector, and again in the 1870s and 1890s; and there were also moments of decline. But the continuous importance throughout the century of the mining industry is very clear. This is the main contribution of this book to the Peruvian and Latin American historiographical debate. Mining also provides, however, a lens through which to explore other issues that have preoccupied other scholars, including questions of internal trade and transportation, business entrepreneurship, labor, and the condition of the Andean peasantry.

The data collected in this book make absolutely transparent the enduring consistency of the mining experience in nineteenth-century Peru. Therefore, in a world-historical context, this South American nation should be added to the list of countries that contributed to the world economic development with mining exports. Peru was one of the first world producers of silver, and also one of the world's largest producers of gold. And since gold and silver constituted the basic currencies, and the monetary standard, for the international economy in the nineteenth century, the world economy depended on this supply of precious metals for its monetary functioning.

However, as this book clearly shows, while the Peruvian mining economy until the 1890s was basically a small-scale economic venture based on the production of high-yielding silver and gold metal ores, in other Latin American countries (Mexico, Chile, Bolivia) different developments were taking place. Nineteenth-century Mexico, for example, was no stranger to the production of iron, and the scale of its silver extraction and processing, while still using the old colonial *patio* system, was on average six times larger than Peru's. Finally, the transition to a more modern mining system—involving large amounts of capital, consolidated and vertically integrated foreign companies, mostly U.S. based, and huge smelting and refining plants—took place in Mexico earlier in the century and at a scale never seen before in all Latin America.[17]

Although the field of comparative studies in mining history is just starting, it could also be argued, based on the evidence provided in this book, that the scale and scope of mining developments in nineteenth-century Chile were also far ahead of those of Peru, particularly from mid-century on. In the 1850s, for example, silver production in Chile was almost twice that of Peru but, more important, Chile was already embarked on the industrially based production of copper. Finally, in Chile the development of coal mining was soon to be another of the industrial results of a mining economy already gaining ground toward heavy industry and large capital investments.[18] In Bolivia, on the other hand, the transition to a more industrial, capital-based, mining sector took place in the 1870s, two decades before a similar transition in Peru, although the Bolivian experience shares with Peru many commonalities in terms of the use of the labor force and the continuities of various productive conditions, whose origins date to colonial times.[19]

Recent historiography of nineteenth-century Peru has been enriched by many new studies. Unfortunately very few of them treat the roles of the mining economy and society. Even more, as I have mentioned earlier, there is a trend in current Latin American historiography to focus on cultural and political issues and neglect, or totally abandon, social and economic themes.

Not long ago, for example, Paul Gootenberg studied the political pressures of various groups to influence the commercial policies of the Peruvian central government in the first half of the nineteenth century. These groups lobbied the government either for the opening of free trade or the building of trade barriers through the action of several military caudillos. Gootenberg identified among the liberal free-traders: the consuls who represented the most important foreign nations trading with Peru (Great Britain, France, the United States), the Bolivarians, the "internationalists, worldly philosophers," and the Arequipa-based "southern secessionists." Among the protectionists were "the agrarian nationalists of the north," Chilean interests, Lima artisans, and, most important of all, the Peruvian merchants represented in the Tribunal del Consulado, the Lima guild of merchants. Despite the title of this very relevant book, *Between Silver and Guano,* mineowners are not mentioned in these trade disputes.[20] Gootenberg has also written important works on price history, Indian peasant demography, and the economic policy debate in which various intellectuals and politicians participated— ranging from the perspectives of protecting "infant industry" or, at the other end of the spectrum, opening liberal free-trade policies.[21]

Christine Hünefeldt has written on the lives of slaves in the first half of the nineteenth century and also on the social living conditions of women and the structure of Limenian families.[22] Peter Blanchard too has dealt with slavery and its abolition in the 1850s, a process that followed a gradual path since the San Martín decrees of 1821 that "gave slaveowners time to adapt."[23] Blanchard's book complements Frederick Bowser's work on colonial Peruvian slavery,[24] and fills the gap between what we knew already about labor in coastal colonial plantations[25] and the more modern and capitalist ones that started to take off after the second half of the nineteenth century.[26] Another recent study of nineteenth-century Peruvian slavery is that of Carlos Aguirre, who contends that the abolition process was principally a result of the actions of the slaves themselves.[27]

Alfonso Quiroz, in a tradition that follows earlier works by Carlos Camprubí, has written a history of finance from the mid-nineteenth to the mid-twentieth centuries and of the role that domestic and foreign capital played in achieving its "visions of development."[28] In a previous work that was also used as input for his "history of finance," Quiroz deals with the relationships between the financial structure and the Peruvian economy between 1884 and 1930 to evaluate the efficiency of financial intermediation mechanisms.[29] Nils Jacobsen has studied a particular region of the Peruvian Andes—Azángaro, in the department of Puno—with respect to the economic and social evolution of its trade and land tenure patterns.[30] Finally, Florencia Mallon wrote not so long ago a thorough study on the rural communities and peasantry of the Peruvian central sierras and how they confronted commerce, industry, and poverty.[31]

This is but a short list of recent books published in the United States dealing with nineteenth-century Peru. And, of course, the contribution of Peruvians writing in Peru has been not only continuous but much more extensive. Previous works of value that present broad interpretations of nineteenth- and even twentieth-century Peruvian history include those of Ernesto Yepes and Julio Cotler, which have been reassessed by the recent historiographical literature.[32] Some reviews, more or less accurate, of the Peruvian historiography are those of Pablo Macera, Fred Bronner, Heraclio Bonilla, José Deustua, Christine Hünefeldt, and Nelson Manrique.[33] It is impossible not to mention here the monumental work of Peruvian historian Jorge Basadre[34] and the many contributions, innovative approaches, and sophisticated analyses of Pablo Macera, who, since the late 1950s, opened such new fields as social and economic history, rural and agrarian studies, as well as sex, language, and culture.[35] More recently he has written several works on the social history of Andean art.[36]

Unfortunately, mining has not been the central question for many of these studies, not even for those that focus more on the economy. There has also been a great tendency to understand nineteenth-century Peru from a rather guano-centric and Lima-centric perspective. Also, for many years nineteenth-century mining was thought of as scarcely significant compared to the well-known, or at least much better studied, mining activity at the apogee of Potosí during the sixteenth and seventeenth centuries.[37] That privileged era produced the European "price revolution"; in comparison, according to historians, mining in nineteenth-century Peru simply did not exist.[38]

Fortunately, historiographical contributions such as that of John Fisher or, later, Enrique Tandeter moved the era of scholarly attention to the late eighteenth century.[39] Both authors emphasized another critical period of mining growth in Potosí and, more surprisingly, in Cerro de Pasco. My study will extend this interest to the nineteenth century, when the mining center of Cerro de Pasco in the central sierras of Peru played a key role in local, regional, and national societies.

Mining in Nineteenth-Century Peru: Propositions and Theoretical Considerations

This book tries to modify the conventional view of nineteenth-century Peruvian history by providing substantial data on the impact of mining on the national society and economy. The conventional view sees five moments in the evolution of the country in the nineteenth century: (1) the crisis of Independence; (2) the post-Independence period, roughly between the 1820s and the 1850s, a period that has been understood as one of political instability and economic decline; (3) the age of guano; (4) the crisis of the War of the Pacific, 1879–1884; and (5) the reconstruction period after the war, a moment of economic recovery that has usually been seen as a result of the agreements of the Peruvian government with international bondholders to write off the Peruvian foreign debt, the so-called Grace Contract.[40]

The revisionist aspect of this book is its focus on the contributions of mining in order to alter or modify this standard historiographical periodization. First, this study demonstrates that mining was an extremely important sector of the national economy that experienced a crisis with the wars of Independence, but recovered very quickly afterward. It contradicts the conventional view that a crisis in mining and in the general economy occurred after Independence. Indeed, mining enjoyed a period of growth in the 1830s, particularly in silver mining in Cerro de Pasco. That growth culminated in a boom in the 1840s.

This book also shows that mining contributed to the development of the national economy and society and did so continuously throughout the century, notwithstanding some fluctuations. Mining's contribution to Peru's gross national product (GNP) throughout the century was large and significant. This share is still difficult to measure fully, but my research should erase any doubt of its significance. However, it is also true that we still do not know the exact evolution of the Peruvian GNP in the nineteenth century, although there have been some attempts to measure it for specific dates. This work should help to enhance those calculations.[41]

As central as mining was to forging a domestic economy and transforming national society, it was also crucial for the export economy, since this sector was the most important component of Peruvian exports before the guano age. Mining also retained some relevance in the export economy during and after the guano boom. In later chapters I return to the relationship between these two critical nineteenth-century export sectors. The force of mining to feed other economic activities, however, was most clearly felt in the domestic arena. Mining, particularly silver mining, was key in the organization and supply of a domestic market in nineteenth-century Peru. Silver metal and coins supplied the market economy, the internal market, with a strategic good that facilitated commercial exchanges. This market economy existed parallel to the natural, rural economy of peasant communities and haciendas. Some pages in this book will discuss the circulation of silver metal in the domestic market and, to a lesser extent, the circulation of money. But mining should not be seen only as a supply sector. In terms of demand, mining centers throughout Peru also consumed goods that mining production required as raw materials and inputs; they also stimulated demand for consumer goods for the resident population and laboring classes. The exchange of these goods, as well as the injection into the economy of silver metal and money, created strong links with the domestic market. Thus, mining set in motion a fluid dynamic that mobilized the domestic economy too, creating, among other things, networks of exchange.[42]

In social terms, the mining sector generated the structural support of a social group that was part of the Peruvian dominant classes: the mining elite. This group of mineowners and business entrepreneurs was part of the Peruvian oligarchy but has not been properly studied, because, among other reasons, Peruvian historiography has focused on the guano elite and the larger group of *hacendados,* the landowners. In this book I only begin to fill this gap by describing the mining elite; I do, however, suggest its importance as a small segment of the Peruvian dominant classes.

At the other end of the social spectrum stand the mine workers, the social group or class responsible for the direct extraction of mineral ores and metals from Peruvian mines. I contend that these workers were part of the large segment of the Andean Peruvian peasantry who migrated from the countryside to work in the mines but never lost their links to their peasant villages and families. They were, then, a labor force constantly in motion, transitional and temporary, not a settled mining proletariat. At least this was the case until the end of the century, when a series of transformations began in various mining centers, particularly in the sierras of Lima and in Cerro de Pasco. These transformations eventually would create a mining proletariat distinct from the peasantry and cut off from the peasant way of life, but this phenomenon would continue well into the twentieth century.

This book also deals with the transportation system that supplied the mines with raw materials and consumer goods. This network also took the minerals and metals out of the mining centers to various regions of the country, and to the ports, to be exported abroad. Until the 1850s, but more properly until the 1870s, when a national railroad system that penetrated the Andean mountains was built, these networks and transport activities were carried out by muleteers working with mule and llama teams and dependent on local and regional merchants. The opposition and conflict between these two systems of transportation, muleteering and the railroad, will give us some clues to understanding the social and economic impact of mining in nineteenth-century Peru.

For the people of the nineteenth century silver was a precious good. It was money; it meant wealth. Thus, mineowners in nineteenth-century Peru were absolutely focused on finding silver ore deposits from which they could extract the precious metal. They were bewitched by silver and they could not perceive that more humble mining goods, such as copper and tin, or even iron and coal, could have also been great sources of wealth. They were so involved in their pursuit of silver that they were unable to create the conditions for alternative sources of development until the last decade of the century. This book, therefore, is also an effort to find answers for this economic paralysis in nineteenth-century Peru.

One perspective that has been used to understand economic matters tries to focus exclusively on measurable indicators of the social and economic behavior of human actors. Demand, supply, the market, production, consumption, employment, are reified as measurable variables that interrelate among themselves. But that reification hides the human dimension contained in each of them. In this perspective, as I stated above, economics tends to be a mathematical science, akin to business administration, not a field of the human or social sciences.[43]

I share, in this sense, many of the ideas discussed in a book edited by economist Warren J. Samuels.[44] The economy is truly a system of power and, as Philip A. Klein puts it, no total impersonal market action exists without some personal decision or state presence, no "pure allocation without valuation."[45] A direct and totally exclusive opposition between the market and the state is, however, misleading. In a precapitalist market the agrarian, rural economy leads the way, following, in the words of British historian Eric J. Hobsbawm, "the fortunes of harvests and seasons."[46] Quite another phenomenon is the market in a capitalist industrial setting, with its large manufacturing companies and wage workers. The role of the state fulfills different functions in these two contexts. But the important matter to consider is the agency of varied social groups, classes, human communities, even individuals, which interact and still counteract the power of the state, the market, the owners of land and capital. Valuation, in this sense, raises the possibility of some human control or action in an otherwise totally alienated context.[47]

Nor is the market merely an economic and social phenomenon. It is also a culture.[48] In Andean countries such as Peru, Bolivia, or Ecuador, there has been a long cultural tradition of nonmarket relations, of reciprocity networks among peasant community members and among peasant communities in larger entities, sometimes called ethnic chiefdoms. This social and cultural tradition is one of the preponderance of kin relationships, of barter, of multiecological organization and exchange.[49] It has also been one of self-sufficiency, of redistribution of goods using different ecological niches and ritual exchanges that were organized by ethnic authorities.[50] In sum, this is a world of peasant economies, agricultural communities, and rural rhythms, with a long pre-Columbian history.[51] It is a world different from and, to some extent, opposed to the market.

The Andean tradition, therefore, has been closer to Chayanov than to Lenin, with its long-lasting existence of peasant communities, *ayllus, parcialidades,* and villages.[52] These communities were not necessarily the closed corporate ones that Eric Wolf once suggested for Mesoamerica.[53] Instead they were relatively open but nevertheless retained a separate identity, culture, and language. I appeal to this very long tradition in the opening epigraph to this book, a verse from a *wayno* (folk song) gathered by researchers in the central sierras of Peru and sung in Quechua Wanka in the nineteenth and still in the twentieth centuries.[54] The other epigraph, from Virgil, denouncing gold as a human evil, is in Latin, to appeal to the Western tradition that the Spaniards brought to the Andes when Pizarro first landed on the coast of Tumbes in late 1531.

The majority of the mine workers in nineteenth-century Peru, whether in Cajamarca, Cerro de Pasco, or Puno, spoke dialects of Quechua or Aymara. They were Andean indigenous peoples that kept strong links to their peasant and rural communities. In this sense, the Andean tradition of the sixteenth century, and even of earlier times, was still alive in the nineteenth century. This tradition, however, was not an immutable social and cultural phenomenon. On the contrary, it has always been changing, adapting to new circumstances. It adapted, for example, to the new labor requirements of the Potosí mining boom in the sixteenth and seventeenth centuries and the reinstallation of *mita,* the forced draft labor that existed during Inca times; it also adapted to the introduction of the market, to market relations, and to a distinctive money economy that started to prevail in Peru since the sixteenth century. Adaptation, accommodation, and resistance have characterized this dynamic Andean tradition.[55]

15

Since the sixteenth century, then, the Andean tradition has been in contact with the market, according to one author, in a process of "de-structuration" and "re-structuration," with the introduction of money, market relations, wage labor, and the forced payment of tribute, sometimes in cash, sometimes in goods or labor.[56] Other authors have written about the creation of several market spaces in the Andean region, and the market effect that the internal circulation of silver money produced in the area.[57] Finally, there is also the reference to a varied price dynamics as a result of different regional conjunctures affected by this Andean way of realizing economic production.[58] In any case my emphasis will be given to market building, a process that has continued in the Andean area from the sixteenth century to today. Note that I say *market* and not *capitalism*. The nineteenth century, in this sense, is a crossroads between this long Andean agrarian tradition of peasant communities and later agricultural haciendas that led the evolution of the national economy, and another, more recent tradition, that started to take place only since the sixteenth century and represented instead the presence of the market, market relations, money, and capital.[59]

One historical phenomenon, then, is the presence of the market in a "natural," rural economy; another is industrial capitalism, with its requirement for continuous and increasing investments of capital. In general terms, Andean peasants' contact with the market since the sixteenth century on has been exploitative because they have not had any control of the market. They even were forced to buy goods at prices set monopolistically by representatives of the state or by merchants. Such was the forced distribution of goods during colonial times (*reparto de mercancías*), and the forced or manipulative buying of wool *(rescate de la lana)*, at the end of the nineteenth century.[60] Thus, the market is not an impersonal, objective phenomenon. It is a personal phenomenon, particularly in agrarian, rural societies. There are people behind the buying and selling of goods and services. There are people behind the labor market. People who hire workers, laborers, people who are being hired, people with their own names and biographies. People embody the market. Therefore, the market expresses a power relation between different social groups in society that have access to, touch, even shake "the invisible hand."

I will try to develop in this book, then, an approach different from modern neoclassic economics. The reader will find information here about the levels of production and exportation of the different mining goods, their prices, the number of mines, of mineowners, and of workers. This is indispensable information to reconstruct the nature of the mining economy in nineteenth-century Peru. Hence this is in part a quantitative and an empirical study. But when confronted with important historical problems, instead of developing abstract mathematical models I have preferred to use simple descriptions, to narrate innocently what the testimonies tell us about the economic life of the nineteenth century. I have also trusted these historical documents as a reliable source to understand the past, instead of adopting a rather postmodern attitude of not believing that they reflect the lives of people in the past, instead, finally, of thinking that history is just another method of textual analysis.

Therefore, although I will also discuss questions of production, consumption, and exchange, I will try to keep my eyes open to the human dimension behind the social economy of the last century. In a serious personal crisis after the defeat of the French army by the German blitzkrieg in the Second World War, French historian Marc Bloch asked himself: "What is it that seems to dictate the intervention of history?" He answered with passion: "It is the appearance of the human element."[61] In my understanding of economics as well, it is the appearance of the human element that makes distinctive and worthwhile the study of economic phenomena. This is why I have used in this book the idea of social economy, and of mining as a social system. The social is not isolated from the economic aspects of human life.

The historian, in the end, is just a good listener. One who reads the documents, gathers oral and cultural traditions, sees the natural

17

and social landscapes of today, and listens to the many voices of the past. If the historian looks for archival documents—the main source of historical studies, according to a venerable tradition—this is just to read between lines the lives of the men and women that really created history. The historian is just a good listener—at any rate, a listener.

Listen to the noise we hear when the Peruvian mines are worked. Listen to the conversation of the workers, almost a whisper, to the commands shouted by the overseers, to the merchants trading silver ingots. Listen as the story begins.

Chapter 2

Mining Production in Nineteenth-Century Peru

Queen Silver, Prince Copper

L'histoire dite économique, en train seulement de se construire, se heurte à des préjugés: elle n'est pas l'histoire noble. L'histoire noble c'est le navire que construisait Lucien Febvre: non pas Jakob Fugger, mais Martin Luther, mais François Rabelais. Noble ou non noble, ou moins noble qu'une autre, l'histoire économique n'en pose pas moins tous les problèmes inhérents à notre métier: elle est l'histoire entière des hommes, regardée d'un certain point de vue.

(The history that is termed economic, which is just being done today, has already run into prejudices. It is not noble history. Noble history is the vessel that Lucien Febvre built; not Jakob Fugger, but Martin Luther, or François Rabelais. Noble or not, or less noble than the others, economic history does nothing less than to raise all the concerns inherent in our profession. It is the entire history of humanity, seen from a particular point of view.)

—Fernand Braudel, *La dynamique du capitalisme*

A REVIEW OF the social and economic historiography of nineteenth-century Peru shows that more interest has been focused on agrarian and rural problems, or on the guano industry as an export sector, than on mining.[1] An explanation for this omission resides in the prevailing idea that the mining economy suffered a tremendous crisis and decline with the end of the colonial regime and the wars of Independence. In 1924, for example, a mining engineer declared that "in the last years of the colonial era, mining was only a shadow of what it was before. . . . When Independence was proclaimed in 1821, the abandonment of mines was generalized due to the emigration of many rich Spaniards who worked in mining; finally, the interior military campaign that ended in December 1824 with the battle of Ayacucho killed the industry because almost all mines in the provinces of Lima and Junín were paralyzed too."[2] A similar comment was repeated recently: "At the end of the eighteenth century, after two hundred years of irrational exploitation, most of the mines were closed. The Indian population was dramatically reduced and the richer mineral veins were exhausted. For mining production, new techniques were needed. Mining did not recover until the end of the nineteenth century."[3] And historian Alberto Flores Galindo affirms that "at the end of the eighteenth century, according to the data of Javier Tord, mining production in Cerro de Pasco was in full swing. With Independence and the subsequent wars (the central sierra was plagued by guerrilla activity), the commercial circuits were interrupted and mining entered a period of full prostration."[4]

This view of post-Independence decline was not shared by nineteenth-century observers, who were more aware of the importance of mining for the Peruvian economy of the nineteenth century. Thus, at the beginning of the nineteenth century, José Morales y Ugalde declared that "the principal wealth of our nation is the metals that its mountains contain."[5] At mid-century the Swiss traveler Johann Jakob von Tschudi noted: "It is surprising the incalculable wealth that has been obtained and it is still obtained from the mines of Peru."[6] And at the end of the century Pedro Venturo commented that "Cerro de Pasco is a privileged place: its mines of coal, silver, copper, gold, lead, zinc, etc., all contribute to assure it a very promising future *[un porvenir muy lisongero].*"[7] By themselves, however, these testimonies are unconvincing. I will attempt to demonstrate their accuracy by reconstructing the quantitative dimensions and the social and economic characteristics of the mining sector in nineteenth-century Peru.

While I am doing this I will also attempt to reflect on the field of social and economic history. I will try to show that the economy in any country, in any period, also has a social dimension that affects ordinary people's lives. In this sense, mathematic economic models to calculate new rates of economic growth, savings, or unemployment could also be translated into empirical experiences for workers, the middle class, or business people, for the elderly, women, children, and so forth. I will start by establishing the basic variables necessary to apprehend the economic development of the Peruvian mining sector in the nineteenth century. I will discuss production levels, volumes of silver, gold, copper, and tin extraction, exports, prices, and commercial and market values. This is, then, the chapter of graphs. I will proceed, however, one step at a time, following an inductive as well as a deductive method. I will use simple mathematical and statistical techniques, tables, and graphs to prove my points, to quantify social realities. Quantification should only be another way of thinking about social problems and of showing how things happened, who is winning and who is losing in any particular society in a specific period of time. My reluctance to separate economics from social analysis has moved me to develop the idea of the social economy of mining in nineteenth-century Peru. The reflection of French historian Fernand Braudel fits perfectly here: "[Economic history] is the entire history of humanity, seen from a particular point of view."[8]

I will start with the precious metals: gold and silver. I will move later to the study of two of the most important industrial metals: copper and tin. I will follow the same framework to analyze both mining sectors, trying to reconstruct their basic economic variables.

The Mining of Precious Metals: A Quantitative History

The most important mining products in nineteenth-century Peru, following a colonial tradition, were the precious metals: gold and silver, but especially silver. The other minerals (copper, tin, coal, lead, etc.) that started to be significant in the late eighteenth and the beginning of the nineteenth centuries never reached the importance of silver. Only at the end of the nineteenth century (1897) did copper start to be competitive with silver, displacing it finally after 1906 as the most important product of Peruvian mining.

Several works refer to silver production in nineteenth-century Peru.[9] A comparison of their data or descriptions with those of earlier authors, whether from the nineteenth century[10] or the beginning of the twentieth,[11] has allowed me to construct a complete record of silver production at the national and regional levels. Of all the regional centers, silver production registered at the official smelting house *(la callana de fundición)* of Cerro de Pasco was the most important, since it was the largest mining center in nineteenth-century Peru.[12]

Of the nineteenth-century authors who produced numerical accounts on mining, the records of the Peruvian savant Mariano Eduardo de Rivero y Ustáriz, who was closest to the mining sector, are the most reliable. During his lifetime Rivero y Ustáriz was general director of mining, the highest government position related to mining, as well as prefect of Junín.[13] The mining center of Cerro de Pasco was thus under his supervision on several occasions during the nineteenth century. It is for this reason that the figures that appear in the *Memorial de ciencias naturales y de industria nacional y extranjera,* the scientific journal he published in the 1820s and 1830s, form the basis for my own table. He later reproduced these figures, with additional data referring to mining production for the following years, in his *Colección de memorias,* published in 1857.[14] For the second part of the nineteenth century, his followers (Mariano Felipe Paz Soldán, Maurice Du Chatenet) are the best sources for obtaining real data on mining production.

Among contemporary authors, it is the economist and economic historian Shane Hunt who has gathered the largest and most reliable sources to reconstruct silver production in Cerro de Pasco since 1830.[15] Hunt is more concerned, however, with discerning a Peruvian export cycle in the nineteenth and twentieth centuries than with mining production itself. Table 2.1 incorporates all these references and methodological observations.

TABLE 2.1
Registered Silver Production in Pasco, 1771–1898
(in marcs of eight ounces)

Year	Production	Year	Production	Year	Production
1771	106,606	1814	192,267	1857	201,207
1772	97,938	1815	156,719	1858	202,825
1773	87,927	1816	176,993	1859	203,445
1774	82,128	1817	145,209	1860	194,435
1775	60,693	1818	167,523	1861	232,854
1776	71,687	1819	190,427	1862	201,500
1777	64,436	1820	312,931	1863	204,493
1778	63,602	1821	—	1864	223,812
1779	77,071	1822	—	1865	175,791
1780	70,366	1823	—	1866	217,228
1781	73,933	1824	—	1867	205,261
1782	69,979	1825	56,971	1868	217,230
1783	72,236	1826	166,118	1869	208,945
1784	68,208	1827	221,501	1870	203,883
1785	73,455	1828	201,325	1871	309,313
1786	109,100	1829	99,835	1872	201,066
1787	101,162	1830	95,261	1873	183,355
1788	120,046	1831	135,134	1874	177,942
1789	121,413	1832	219,378	1875	169,679
1790	117,996	1833	257,069	1876	169,878
1791	123,789	1834	267,126	1877	178,449
1792	183,598	1835	276,744	1878	159,630
1793	234,942	1836	244,404	1879	169,185

TABLE 2.1 (continued)

Year	Production	Year	Production	Year	Production
1794	291,253	1837	235,856	1880	130,049
1795	279,621	1838	251,932	1881	105,515
1796	277,553	1839	279,620	1882	104,470
1797	242,948	1840	307,213	1883	119,828
1798	271,861	1841	356,118	1884	125,926
1799	228,356	1842	387,919	1885	131,497
1800	281,481	1843	325,458	1886	146,500
1801	237,435	1844	274,602	1887	137,200
1802	263,906	1845	251,039	1888	131,900
1803	283,191	1846	281,011	1889	161,700
1804	320,508	1847	245,307	1890	165,600
1805	306,050	1848	272,994	1891	159,600
1806	161,193	1849	229,889	1892	163,000
1807	242,031	1850	219,548	1893	166,500
1808	243,295	1851	235,702	1894	148,800
1809	285,731	1852	218,558	1895	174,900
1810	240,220	1853	238,423	1896	172,735
1811	251,317	1854	202,695	1897	158,781
1812	180,061	1855	257,928	1898	125,088
1813	180,897	1856	218,356		

Sources: Memorial de ciencias naturales y de industria nacional y extranjera 1.4 (Lima, March 1828): 164; Rivero y Ustáriz, *Colección de memorias,* 1:219-20; Paz Soldán, *Diccionario geográfico,* 208-9; Hunt, *Price and Quantum Estimates,* 51, table 19. See also Archivo General de la Nación, Lima (hereafter AGN); Sección Histórica del Ministerio de Hacienda (hereafter SHMH), OL 186, caja 117, ff. 652-61. Some other annual figures lightly different appear in the Lima newspaper *El comercio,* 19 May 1839, 31 December 1855, 3 January 1856, 8 January 1857, 14 January 1858, and 15 January 1859; and in the *Accounts and Papers* of the *British Parliamentary Papers,* vol. 64, pp. 208-9, 1847 (Public Record Office [hereafter PRO], Foreign Office [hereafter FO], London). See also Carlos Camprubí: *Historia de los bancos en el Perú (1860-1879)* (Lima: Talleres Gráficos de la Editorial Lumen, 1957), 169; Macera, *Estadísticas históricas,* 74-77 and 104; Heraclio Bonilla: *Gran Bretaña y el Perú: Los mecanismos de un control económico* (Lima: Instituto de Estudios Peruanos and Fondo del Libro del Banco Industrial, 1977), 5:183; Fisher, *Minas y mineros,* 243-44 and appendix; McArver, *Mining and Diplomacy,* 294, appendix A; Tarnawiecki, *Crisis y desnacionalización,* 84, table A1; and some reports of the British consuls in Peru in Heraclio Bonilla, ed., *Gran Bretaña y el Perú, 1828-1919: Informes de los cónsules británicos,* 4 vols. (Lima: Instituto de Estudios Peruanos and Fondo del Libro del Banco Industrial, 1975), 1:184, 258, 303.

I have deliberately included data from part of the eighteenth and all of the nineteenth centuries to demonstrate cycles of mining production, especially in Cerro de Paso. Hunt's work also provides a first step for reconstructing mining production at the national level. Nevertheless it is important to consider primary sources too. Until 1834 we have the records of the seven official smelting houses that operated in nineteenth-century Peru: Pasco, Lima, Trujillo, Huamanga, Arequipa, Tacna, and Puno. These data have been used before by authors such as Mariano de Rivero in the nineteenth century or John Fisher in the twentieth.[16] Since then mining data are more scattered and incomplete: some appear in the *Boletines oficiales de minas y petróleo* or in the *Memorias* of the ministers of finance, but these are not enough to construct a complete set. To obtain national figures then, we have to make estimates for some years, relying mostly on Pasco records (the most reliable data throughout the century), a procedure that Hunt also used.[17] Based on table 2.1 and on these considerations, I have developed the following graph, which presents the output of silver mining at the national level and in Cerro de Pasco from 1771 to 1898.

The graph shows that there were three great moments of growth in Peruvian silver mining production. Between 1790 and 1800 mining production at the national level reached more than half a million marcs, some 220,000 kilograms, per year. A second moment of growth occurred around the 1840s; although not comparable to the first at the national level, it was outstanding in Cerro de Pasco. This moment of mining growth in Cerro de Pasco was the most important one in that area of the country during the whole century. Finally, there was a third moment of growth at the end of the century, mostly at the national level, whereas for Cerro de Pasco this was a period of economic decline.

Both curves, however, those of national and of Pasco production, are very similar. The oscillations of their growth and decline are nearly identical. The years of the wars of Independence, for example, represented a complete breakdown at both levels. This extreme crisis, which at some point totally paralyzed mining production, could explain some of the references mentioned above whose authors believed that the momentary crisis of the years of Independence represented an enduring trend.

The similarity between the two curves also shows that silver production in Cerro de Pasco was the main source for the national output. During colonial times, as we can see in fig. 2.1, the national output was almost double that of Cerro de Pasco. This means that the contribution of other mining areas, such as Hualgayoc or the

FIGURE 2.1
Registered Silver Production in Peru and Pasco, 1771–1898

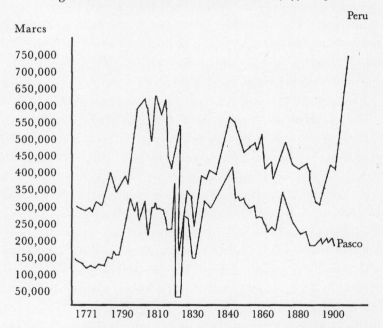

Sources: Table 2.1; Rivero y Ustáriz, *Colección de memorias,* 1:225–26; Fisher, *Minas y mineros,* 243–44; and Hunt, *Price and Quantum Estimates,* 57–59, table 21. See also Deustua, *Minería peruana.* 243–61; Contreras, *Mineros y campesinos,* 50–54; and Deustua, "Mines, monnaie," chap. 1.

Lima sierras (Casapalca, Morococha), was as important as that of Cerro de Pasco. But later on, after Independence, the national figure closely approximates that of Cerro de Pasco alone, reflecting a centralization process and growing dependence on the Cerro de Pasco production to consolidate national results. The two curves diverge again, during and after the War of the Pacific (1879–1884), indicating that the other mining centers again contributed significantly to the national output.

Four economic cycles may be distinguished, each with its corresponding moments of growth, apogee, crisis, decline, and recovery:

The first cycle, between 1771 and 1821, was the last moment of colonial mining and has already been studied by John Fisher.[18]

The second cycle, between 1821 and about 1860, had its apex in 1842, when Cerro de Pasco silver production reached its highest point for the whole century: 387,919 marcs of pure silver were extracted from its mines.

The third cycle took place between 1860 and 1882, that is, in the twenty or so years leading up to the War of the Pacific. The disruptions that the war set off ended this cycle and engendered a new crisis in the mining economy. During the war, and especially after 1881, the capital of the country, Lima, as well as the largest mining center, Cerro de Pasco, were battlegrounds that were later occupied by the Chilean army.[19]

The end of the war brought a new process of recovery and growth in mining production that was felt much more intensively at the national level than at the Cerro de Pasco regional level. This fourth cycle brings us to the twentieth century. Peruvian mining had its highest point of production, and the highest point of production for the whole century at the national level, in 1898, when Peruvian mines produced 717,475 marcs of silver.

But mining production, economic output, or the amount of silver extracted in a given year represent only one dimension—the "real sector"—of a variety economic realities. For an entry into the social economy of mining, we need to look at silver prices—the market value at which a good is exchanged, a market value, finally, that is fixed in terms of money.

The evolution of silver prices in the Peruvian economy over the *longue* or *moyenne duréekem*[20] presents a different picture of the mining sector than that portrayed by production data. Figure 2.2 shows the evolution of international silver prices from 1833 to 1898 on the London commodities market (in pence per pound). London was at the time the most developed international market for silver and other commodities, followed closely, as the century progressed, by Paris, Hamburg, and New York. The graph also traces the volume of silver production (in marcs) in Peru during these years, according to the data previously discussed. Finally, a time series of the market value of silver in nineteenth-century Peru is also included. This last curve was obtained by multiplying the volume of silver production by its price, the result expressed in British pounds sterling.

Silver prices, particularly on the London market, were stable for much of the century, especially until the 1850s, and even experienced a slight growth in the 1860s, when they reached more than 62d. per ounce. This was the time of the second Peruvian mining cycle (1821–1860), the most important one for the whole century, although not as lengthy nor as significant as the late-colonial cycle (1771–1821). Silver prices, however, dropped after the 1870s, from an average of 61d. per ounce to 42d. in 1888, 29d. in 1894, and 26d. in 1898. This was a dramatic price crisis for silver at the world level, and obviously affected Peruvian mining. And since Peruvian silver production did not increase dramatically in the 1870s, the decline in prices sharply affected its total market value.

Figure 2.2 demonstrates the importance of silver mining throughout the century. The average annual market value of silver produced between 1833 and 1898 was around £670,000, although this continuous production had its fluctuations and cycles, with several moments of boom and bust. The peaks of 1842, 1849, 1851, 1871, and 1898 were

FIGURE 2.2

Volume, Price, and Value of Silver Production in Peru, 1833–1898

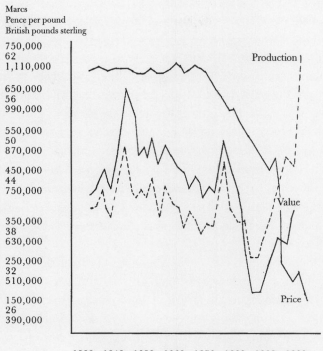

Marcs
Pence per pound
British pounds sterling

Production volumes in marcs; prices in pence per pound; value in pounds sterling.

Sources: The silver production curve is based on table 2.1 and fig. 2.1. Prices are based on "Statement Showing the Annual Average Price of Bar Silver per Ounce Standard" of the "Statistical Tables Showing the Progress of British Trade and Production, London, 1896," PRO, FO, *Accounts and Papers* of the *British Parliamentary Papers,* 1896, vol. 76, p. 48. See also the price series from 1835 to 1915 of Carlos Jiménez, *Estadística Minera en 1915* (Lima: Cuerpo de Ingenieros de Minas del Perú, 1916, boletín n. 83); and *Boletín del Cuerpo de Ingenieros de Minas del Perú,* n. 14 (Lima, 1903), 28, which closely follows the British series. I have given the production values in pounds sterling and not in Peruvian soles or pesos due to the great stability of the British currency throughout the century.

especially notable. The two highest came in 1842 and 1871. In 1842 silver values reached £1,019,527, more than 5 million Peruvian pesos, whereas in 1871, silver values still represented some £914,911. After 1871, however, silver values declined steadily, to recover slightly at the end of the century.

The market value of silver, therefore, was substantial throughout the nineteenth century, although average levels declined slowly after the 1840s, but most drastically after the 1870s, due basically to the international crisis in the price of the white metal.

The response to the decline in international prices was slow. Not until the end of the century was more silver produced, but increased output did not compensate fully for falling prices. This slow reaction in production to an international price crisis, which has much to do with the lack of production capacity, will be discussed in the following chapters. For the moment let us come back to the information that figure 2.2 provides.

The output of Peruvian silver mining declined slowly in value throughout the century, with some occasional peaks and some moments of enthusiasm. However, in stark contrast to the traditional view of a crisis in mining at the beginning of the century, particularly during and after the wars of Independence, with a slow recuperation by the end, we find that the first half of the century was brighter than the second. At the regional level we find a similar trend (see also fig. 2.1). Silver production in Cerro de Pasco reached its peak in the 1840s, when international prices for silver were stable or slightly increasing. At the end of the century (after 1893), we observe a new rise in silver production. But prices remained low and mining in the area was visibly shifting to the extraction and refining of a new mineral: copper.[21]

The importance of silver mining for the national economy is in any case incontestable. In spite of its slow decline in terms of value, as a whole the sector produced on average well above £500,000 per year, or more than 3 million Peruvian pesos, a significant amount of wealth.[22]

Silver production revenues sometimes even reached £1,000,000, as in 1842. Therefore it is not surprising that Peru was one of the largest silver-producing countries in the world. Over the course of the century Peru produced around 10 percent of all the world's silver, 14 percent in the first half of the century, when the United States was not yet a major producer. In Latin America, Peru produced more silver than any country except Mexico. Peru was well ahead of Chile and Bolivia in the first half of the century, although in the second half Peru's performance remained stable, whereas Chile and Bolivia increased their silver output (see table 2.2).

The 1860s is then the turning point for Peru's silver mining. Peru could not keep pace with the growth of silver production in other Latin American countries, nor with the United States' upsurge, nor with the great increase in world production. The 1860s also represented a transition for the Peruvian silver economy, from the second (1821–1860) to the third Peruvian mining cycle (1860–1882). Furthermore, each of the three successive Peruvian silver mining cycles was smaller than its predecessor (see fig. 2.1). Table 2.2 shows, for example, that 39 million ounces of silver were produced in the first decade of the century, a level not reached again until the 1890s—although the table also shows that there were two other increases in silver production, in the 1840s (34 million ounces) and in the 1870s (28 million ounces). The transformation of Peruvian silver mining, with the beginnings of its industrial production, occurred only at the end of the century, marking a fourth cycle (1883–1900 and beyond).

One major reason for this "contained development," for the failure of silver mining to take off in the 1860s or the 1870s, was the guano boom. Guano, since 1847 the most dynamic and lucrative export sector, diverted resources and attention from silver mining in nineteenth-century Peru.[23]

Comparison with the performance of other Latin American countries, nonetheless, is complex. Chile surpassed Peruvian silver production in the 1850s, but it was not until the 1870s that Bolivia surpassed

TABLE 2.2
Largest Silver Producers in the World, 1801–1900
(millions of ounces)

Decade	World	United States	Mexico	Bolivia	Chile	Peru
1801–1810	287	—	n.d.	n.d.	2	39
1811–1820	174	—	60	n.d.	4	32
1821–1830	148	—	80	21	7	13
1831–1840	192	—	85	21	12	29
1841–1850	250	—	95	16	14	34
1851–1860	285	0.5	140	21	43	27
1861–1870	390	78	160	22	36	25
1871–1880	668	279	190	44	43	28
1881–1890	972	414	270	88	48	24
1891–1900	1,614	567	480	132	52	44

n.d.= no data

Sources: On Mexican, United States, and world production, see Herbert Bratter, *The Silver Market* (Washington, D.C.: Government Printing Office, 1932), 66–67; Pierre Vilar, *Or et monnaie dans l'histoire* (Paris: Flammarion, 1974), annexe 2, 431–33; Antonio Mitre, *Los patriarcas de la plata: Estructura socioeconómica de la minería boliviana en el siglo XIX* (Lima: Instituto de Estudios Peruanos, 1981), app. 2, 195. On Bolivian silver production (apart from Mitre, *Patriarcas*) see Antonio Mitre, "Economic and Social Structure of Silver Mining in Nineteenth-Century Bolivia" (Ph.D. dissertation, Columbia University, 278–79); and Herbert S. Klein, *Bolivia: The Evolution of a Multi-Ethnic Society* (New York: Oxford University Press, 1982), table 2, 298–99. On Chile, see Pierre Vayssière, *Un siècle de capitalisme minier au Chili, 1830–1930* (Toulouse: Centre National de la Recherche Scientifique, 1980), 112. On Peru, see the same sources as in fig. 2.1.

Peru.[24] In the 1840s, however, Peru produced more silver than both of these countries combined. Even if Chile surpassed Peru in the 1850s, the former did not experience a real mining upsurge either: in the 1860s its production declined compared with the previous decade. Mexico's silver production surged in the 1850s, but from then until the 1890s mining did not show impressive growth. Peruvian silver mining performance was also not impressive, but it was continuous and steady. What accounts for the lack of dynamism in Peruvian silver mining in the second half of the nineteenth century? Why did it not start to grow as in other Latin American countries? Before offering some answers to these and other questions, it is well to remember that silver was not the only precious metal being mined. There was also gold.

The statistics on gold production are less consistent than those of silver, an indication of the relatively greater importance of silver. Nonetheless the data are sufficient to provide us with an image of the evolution of gold mining in nineteenth-century Peru. In constructing figure 2.3 I have used the most consistent information on the production of gold and the minting of gold coins over the course of the century. As we can see there is a period of growth in the minting of gold coins in the 1810s, but it is followed by a decline related to the wars of Independence. The 1840s show a new period of growth, but keep in mind that these are now statistics of production, so they should obviously be higher than the figures for minting, because not all the gold extracted from Peruvian mines was minted. After the 1850s gold production tended to decline.

There are some differences, nevertheless, if we compare my numbers with those of Macera and Jiménez.[25] For those researchers, the 1830s in gold mining was a period of growth, with production reaching 750 kilograms per year on average, compared to 320 kilograms in the 1820s. In contrast, the minting data that I have gathered suggest that the 1830s was a period of decline. Is there a big gap between production and minting in nineteenth-century Peru or are we confronting a problem of research sources? Minting statistics are on average four times lower than Macera and Jiménez's data, and seven times lower than Hunt's data for the same period.[26] Clearly, minting was well below gold production in nineteenth-century Peru.

Hence it appears that significant quantities of gold were produced and that only a portion of this gold was minted and circulated as money. Gold ore was also transformed into metal and circulated in ingots, or as *chafalonía* (crafted pieces of gold). In 1821, for example, a report from the State Customs of Lima stated that gold was circulating in the

FIGURE 2.3
Production and Minting of Gold in Peru, 1800–1900
(annual averages for five-year periods in kilograms)

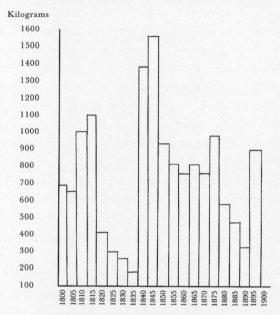

Gold Minting 1800–1840; Gold Production 1840–1900

Sources: For the period 1800–1839 I have used data from the records of the several Peruvian minting houses that appear in John R. Fisher, *Government and Society in Colonial Peru: The Intendant System, 1784–1814* (London: Athlone Press, 1970), app. 3; Manuel Moreyra y Paz Soldán, *La moneda colonial en el Perú: Capítulos de su historia* (Lima: Banco Central de Reserva del Perú, 1980), anexo 1; Alberto Flores Galindo, *Aristocracia y plebe, Lima 1760–1830* (Lima: Mosca Azul Editores, 1984), anexo 6, 253; and "Return of the Number of Marcs of Gold coined at the Mints of Peru," in *Accounts and Papers* of the *British Parliamentary Papers* 64 (1847): 207, PRO, FO. See also José Deustua, "El ciclo interno de la producción del oro en el tránsito de la economía colonial a la republicana: Perú, 1800–1840," *Hisla* no. 3 (Revista Latinoamericana de Historia Económica y Social, Lima, 1984): 23–49, esp. table 2, p. 37, where there is a discussion of these figures and sources. For the period 1847–1900 I have used data from Adolph Soetbeer, *Edelmetall-Produktion* (Gotha: Justus Perthes, 1879), 65–70; and Hunt, *Price and Quantum Estimates,* table 21, 57–59, whose sources are reports of the Ministerio de Hacienda (Lima) and Superintendencia de Aduanas, *Estadística general del comercio exterior del Perú* (Lima, 1897–1900). Wherever possible, I have calculated the five-year averages from real figures. If this was not possible I have used estimations from Hunt's work or my own.

34

city *"en pasta, labrado y amonedado"* (in bullion, wrought, and as coins) and that 3 percent of the value (in pesos) of this gold was charged at the time of its exportation.[27] Thus it is likely that significant amounts of gold were produced in the nineteenth century, although we do not know the exact total for each year. However, based on the years we can account for, I estimate the average annual gold production for the century at 484 kilograms. Of course we should expect that gold mining, as in the case of silver, fluctuated throughout the century. If the annual average production of gold was 484 kilograms, this would account for 5 percent of world production before the discovery of gold in California. By the end of the century the Peruvian contribution to world gold production declined to less than one percent of the world total.[28]

The erratic statistics on gold production reflect the manner in which it was produced in this preindustrial and prestatistical era of the Peruvian economy. The final product, gold metal, was and is a commodity of great value, even in small amounts. Every new discovery, every new mine opened, changed the characteristics of its trade and the characteristics of the whole gold-mining sector, making data collection during the period chaotic. Gold was also a commodity that was easy to transport, again because of the great market value even of small amounts. Finally, gold was sought after as a symbol of wealth, a commodity to be hoarded—more so than silver, which was used mostly as currency. Therefore gold, as a mining good, appeared erratically in the market, in the minting houses, and at the government offices that kept the records of its production or exportation.

The geographical and regional location of gold mines shifted significantly over the course of the century. Huamanga had been the colonial center of gold mining but experienced an enduring decline during the nineteenth century.[29] The sierras of La Libertad (Pataz, Parcoy, etc.), by contrast, kept up production at a constant level,[30] whereas in Puno after mid-century there was a period of growth as a result of the discoveries of gold veins "on the beaches of the Inambiri River" or "in the Challuna valley on the opposite shore of the Guari-Guari River."[31]

35

Archival evidence, as well as data presented in fig. 2.3, suggests that there was a retreat from gold production at the beginning of the national period, between 1820 and 1840. In 1825 there are some indications of this decline in a report from the mining centers of Queropalca and Chuquibamba, and from a report of the same year from Pataz.[32] In 1828 there were complaints that Huallanca "currently is in deep decay *[en suma decadencia]*." And almost twenty years later, in 1846, in Parcoy, department of La Libertad, observers asserted that "the mining industry is declining very fast."[33] It is true that these testimonies could simply be expressing the dissatisfaction mineowners felt with the situation in the country after the wars of Independence, and not a real decline in mining production. But if to these testimonies we add the mint figures presented above, the crisis that gold mining experienced at the beginning of Peru's national period should be more clear. In any case it is evident that the wars of Independence caused several disruptions in the organization of gold production and trade, to which the qualitative archival documentation previously mentioned also referred.[34]

It is only at the end of the nineteenth century that we can properly speak of a recovery in gold production. In 1890, for example, Sir C. E. Mansfield, the British consul in Peru, taking as a source "reports from the Mining School of Lima," drew a very detailed map of the gold deposits in the country,[35] although the document includes prospective mines as well as existing ones. In any case it represents a clear example of an invitation to foreign companies to invest in gold mining, an expanding activity in various regions of the country, particularly in Puno. In Carabaya, in the same years covered by the Mansfield report, the British Inca Mining Company and the Chaquimayo and Inambari Company were already in full production.[36] By 1897 the Peruvian engineer Teodorico Olaechea was already developing a precise estimate of national gold production, taking into consideration the different regions of the country (see table 2.3).

According to Olaechea's data a total of 839 kilograms of gold,

TABLE 2.3
Gold Production in Peru, 1897

Place of Origin	Weight of Raw Gold Kg.	Weight of Fine Gold Kg.
Department of Puno (mines of Sandia and Carabaya)	459.2	—
Regions of Northern and Central Peru	—	174.2
Department of Junín (Chuquitambo mines)	41.3	—
Office of Casapalca	—	43.1
Others*	75.6	45.7
Total	576.2	263

*Includes the regions of Caylloma, Ayabaca, Arequipa, Ayacucho, Huancavelica, and Cusco, each of which produced less than 25 kg. per year.
Source: Teodorico Olaechea, Apuntes sobre la minería en el Perú (Lima: Imprenta de la Escuela de Ingenieros, 1898), 20–21.

whether raw or fine, was produced in Peru in 1897. Nevertheless, he describes, in addition "not a small amount of gold that runs through regional fairs, clandestine exploitation, and exports," which would increase the national total to at least 1,000 kilograms.[37] In 1900, for which there are better statistics, the total amount of gold produced in the nation was 1,633 kilograms. As was true of silver, then, the 1890s witnessed growth in gold production.[38]

Prices for gold have a remarkable stability throughout the century, fluctuating very slightly on the London market between £3 17s. 1d. per ounce in 1811–1820 to £3.18.9 in 1866–1870.[39] The apex was reached on the London market between 1881 and 1885, when gold prices were £3.19.7. Between 1891 and 1900 gold dropped to £3.12.3, the lowest price of the nineteenth century. No dramatic price crisis occurred, as it did in the case of silver, despite the great transformations that the global market of world production experienced during

37

the century with the discovery of gold in California, Australia, and later South Africa.[40] This relative price stability meant that gold mining kept an economic relevance for Peru as a whole throughout the century, although not as great as silver mining. If around the 1830s the commercial value of gold production was £89,339, toward 1860 this commercial value was £103,083, and £119,565 in 1897.[41] In nineteenth-century Peru the value of gold production was on average merely 12 percent that of silver (although the amount of metal produced was only 0.7 percent that of silver). The difference between value and amount was obviously due to the higher commercial value and, therefore, higher prices of gold compared to silver. Gold prices were between fifteen and twenty-one times higher than those of silver throughout the century.

In sum, gold was also a significant sector in the nineteenth-century Peruvian mining economy. Its production levels and market values were smaller than those of silver, but gold production was of substantial importance and less prone to fluctuations and crisis than silver production.

The Mining of Industrial Metals: Copper and Tin

Although the precious-metals sector (gold and silver) figured prominently in the mining economy of nineteenth-century Peru, the production of industrial minerals and metals, such as tin, and especially copper, was also significant for that economy at the beginning of the national period. The economic history of copper shows how it passed from a secondary sector of the mining industry in the nineteenth century to become the most important at the turn of the twentieth, eclipsing the predominance of silver. The statistical data I gathered on copper production are, nevertheless, not as good as those on silver. The paucity of the data reflects again the lesser importance of copper compared to silver or gold.

The Peruvian historian Heraclio Bonilla, in his doctoral thesis and in an earlier article on the "commercial conjuncture" in nineteenth-century Peru, has reconstructed the exports of copper to Great Britain and France for the period between 1833 and 1894 in constant values, not mentioning its nominal values.[42] His research gives us the clues necessary to uncover the trends and cycles of the evolution of Peruvian copper exports.[43] The work of Bonilla should be considered along with that of Shane Hunt, who criticizes some of Bonilla's results. Hunt is mainly interested in following the trends and cycles of copper exports, without considering the regular, year by year evolution of copper production. His figures have been obtained by deflating copper values with a price index at the moment of its exportation.[44] They are, from my vantage point, however, too high, so probably there has been a problem in the conversion from values to volume. In 1839, for example, according to Hunt's figures, 2,611 metric tons of copper (including both ore and refined copper) were exported from Peru to Great Britain, France, and the United States. This amount, some 11,488,400 marcs, was twenty-seven times the amount of silver produced that year (422,840 marcs). Furthermore, it represented only a fourth of the copper exported in 1906, by which time the complex modern technology employed by U.S. giant Cerro de Pasco Copper Corporation produced massive amounts of the metal.

Hunt recognizes that his figures are in general higher than those of Bonilla, stating that the gap is largest between 1833 and 1839, when the ratio between the two sets of data reaches 1:13, and between 1840 and 1849, when it reaches 1:6.[45] Two other reasons for the higher figures Hunt obtains could be, according to his own methodological observations, that in addition to Peruvian exports to (rather than all imports into) Great Britain and France, Hunt includes exports to the United States, Germany, and, to some extent, Chile; and that the ships sailing to all these places under a Peruvian flag could also have been carrying copper from Bolivia or Chile.[46] In contrast to this reasoning and to defend Hunt's figures, I would add that in Peru veins of copper ore are more abundant than those of silver. In fact, the proportion in which they are found today is nine to one in favor of copper.[47]

In any case I have also found some primary documentation on copper production and exports in nineteenth-century Peru. Some of this primary material was processed in Bonilla's work.[48] These data concern copper exports to Great Britain, in volume and value, broken down into: (a) regulus copper from 1851 to 1919, (b) copper ore in the same years, and (c) partially wrought and unwrought copper between 1832 and 1919.[49] These materials, in addition to my own archival sources, render a reasonably detailed, reliable picture of Peruvian copper exports in the last century. The reader should keep in mind that these figures are for export; figures for production are sketchier. The data, divided into exports of copper ore and of refined copper, are presented in figure 2.4.

Similar trends in the evolution of exports of copper ore and refined copper are visible in the graph. Whereas the export volumes remained low in the 1830s and 1840s, there is a tendency toward growth in the following next two decades (1850s and 1860s). The decline at the end of the 1860s and crisis of 1871 were abruptly reversed. In 1878, for example, exports reached 8,131 tons of copper ore and 5,948 tons of refined copper, the highest exports for both goods until then. A new crisis then occurred, obviously related to the War of the Pacific. But a new revival at the end of the century foreshadowed copper's dominance over the Peruvian mining economy at the onset of the twentieth century. In this new period of growth, exports of copper ore reached the highest figures for the whole century: 8,562 tons in 1899 and 12,260 tons in 1900. These were two big leaps forward, well above the century's averages, that indicated the revolution in production copper was experiencing at that time.

Another phenomenon visible in the graph is that the exports of refined copper were consistently lower than those of copper ore. The reason was that refining copper required an industrial process that the Peruvian economy was not still capable of fully developing on a large

FIGURE 2.4

Peruvian Exports of Copper Ore and Refined Copper, 1830–1900

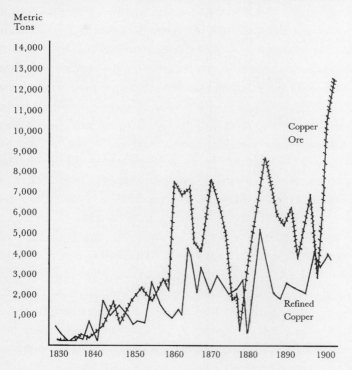

Sources: Bonilla, "Coyuntura comercial," table 5; Hunt, *Price and Quantum Estimates*, 38, table 14; Bonilla, *Mecanismos de un control*, 175–80, tables 12, 13, and 14; and "Foreign and Colonial Merchandise Imported into the United Kingdom from Peru," in *Accounts and Papers* of the *British Parliamentary Papers* 39 (1849): 372, PRO, FO.

scale. Copper was exported mainly as a raw material. But not a raw material as we understand the concept today—goods that have been subjected to processing or packaging. Nineteenth-century Peruvian copper exported as ore was extracted directly from the mines and shipped without any processing whatsoever. These were very heavy raw ores that also contained noncommercial minerals. This fact gives us some indication of the undeveloped nature of the copper mining sector, despite its linkage with the production of an essential modern industrial good: copper. This preindustrial backwardness observed in copper mining was also one of the characteristics of the nineteenth-century Peruvian mining economy, outside the precious-metals sector.

The data presented above show exports of copper only after 1830, as if there were none before then. My primary documentation, however, shows that copper was produced and exported long before 1830. In 1821, for example, the year in which Peru declared its independence, Juan Antonio Gordillo, a Lima customs official, found in the navy stores in Callao 87 copper bars weighing 163 quintals, 47 pounds, that were going to be exported "with the destination Spain."[50] In 1826 Juan Manuel Pinelo y Torres issued a report from Chumbivilcas saying that in that locality a mine for silver, gold, and copper was in production, although it was experiencing some difficulties.[51] In the same year Mariano de Rivero wrote of the existence of mines for silver, copper, iron, and lead in the province of Huacullani, southeast of Puno.[52] Similarly, between May 1828 and December 1829, the Lima mint was buying 143 quintals, 53 pounds, of copper at prices that fluctuated from 62 to 64 pesos per quintal.[53] Hence copper was already produced and exported before 1830, and the Lima mint, as in the cases of silver and gold, was one of its main domestic purchasers.

I have also found colonial exports of copper mentioned in a report by Lima State Customs (la Aduana del Estado de Lima). According to this document, among the colonial products and manufactures "exported to the ports of Spain," one could find not only gold and silver, but also copper and tin. After 1815 copper and tin were taxed at one peso per quintal exported, according to the *Arbitrio temporal.*[54]

There was clearly an internal production and an internal consumption of copper, apart from the export trends shown in figure 2.4. Copper was used during colonial times, and obviously during the nineteenth century, for the fabrication of kitchenware (pots, kettles, etc.) and tools.[55] In 1821, for example, an inventory of the goods belonging to the state in the administration of Cerro de Pasco *(de los efectos pertenecientes al Estado en la Administración del Cerro de Pasco)* mentioned two copper cups *(dos tasas de cobre)* that were produced locally.[56] Copper ore, then, was processed, refined, and worked in various ways through craft labor in smith's shops that formed part of the Peruvian domestic economy. In this way mining became linked to small-scale industry or artisanal work that took place in many areas of the countryside, in small towns, and even in some larger cities, such as Cerro de Pasco.[57]

According to reports of the Lima mint and other sources there were several kinds of copper products: copper in bars, copper *en granella* (in plates), *cobre fino* (fine copper), *cobre refino en granella* (extra fine copper in plates), copper ore, impure metal regulus copper, unwrought and partially wrought copper, and so forth.[58] What did not exist during much of the nineteenth century was a connection between copper mining and a modern industrial sector that transformed the ore into a finished industrial good on a mass scale. This technologically advanced economic sector only began to develop at the end of the century with the building of several sophisticated, capital-intensive smelting plants. Before that, copper ore was worked only in blacksmith shops in small cities and rural areas where specialized craftsmen were concentrated. It was a small-scale industry producing copper intended for internal or local consumption, while *platería* (silver craft work) was a more specialized business oriented toward the consumption of the upper classes.[59]

Industrial processing would have required a whole economic infrastructure, capital investments, the building of industrial plants, transportation by rail, and an industrial demand that were clearly lacking before the discovery of electric power and its transmission through copper wires. However, in neighboring Chile, copper mining had begun to grow in the 1830s to satisfy an increasing industrial demand overseas.[60]

The limited commercial demand for copper was reflected in its prices. The price of the copper bars found in the navy stores in Callao was 22 pesos per quintal; the mercury and gold also found there were 50 pesos and 27,526 pesos per quintal, respectively.[61] Furthermore, the copper that reached the Lima mint had higher prices, between 62 and 64 pesos per quintal, than regular copper ore or metal. But it is also true that the prices of other nonprecious minerals were well above those of copper. The best prices for mercury, for example, reached 125 pesos per quintal, while bars of gold and silver were purchased at prices as high as 26,312 and 1,619 pesos per quintal, or 130 and 8 pesos per marc.[62] Hunt has also compared the prices of copper, silver, and gold, but for a later period, around 1870, when copper enjoyed a larger industrial demand. In any case his data also show that the price of copper was well below that of silver and gold.

In 1821 the ratio between gold and copper prices, according to the report from the navy stores in Callao, was 1,251:1. Around 1870, when gold prices increased slightly and silver prices started to drop, the ratio of gold to copper was 4,504:1 and silver to copper, 298:1. Precious metals then always retained a very high value compared to the price of copper, but in addition copper prices gained more importance in the second part of the century, when the demand for copper, as an industrial metal, also began to increase, as did its production in Peru and the world. As we have seen in figure 2.4, Peruvian exports of copper gained strength in the second half of the century.

With no great commercial demand and a very low market value (as reflected in its low prices), we would expect little production and almost no exports of copper in the nineteenth century. But, as we have seen, that was not the case. The reason for this apparent paradox is that most of the time copper mining was not an independent activity. I have found several references that indicate it was a secondary activity linked to silver and gold mining. In the Peruvian mountains copper veins existed side by side with silver or gold deposits. Therefore, when a mining operation extracted silver ores it was also extracting copper ores at the same time, sometimes in larger amounts. Usually this copper-bearing debris was simply discarded, although a small amount of the copper was often extracted and processed. In these cases the process of extracting copper was a zero-cost operation, because its costs were covered under the silver mining work. The only commercial costs for copper production were, consequently, the processing of copper ore or its transportation to ports and marketplaces. Silver mining and gold mining therefore lowered the operating costs of copper mining.

That was clearly true in 1826 of the metal ores of the Pomasi mines in Lampa, Puno, one of the more productive and wealthier mining areas for silver in nineteenth-century Peru. In this mining center the silver ores contained large amounts of copper, as copper pyrites and other sulphurated copper. The metallic ore of the Nuestra Señora de la O, Chinquiquirá, Descubridora, Copacabana, Trinidad, and other mines in Pasco also contained in 1828 silver-bearing iron oxides, copper pyrites, and *pintas de pavonado* (bluish silver-bearing stains). Similarly, according to Peruvian scientist Mariano de Rivero, exploitation of the mountain of Chuquitambo in Cerro de Pasco for gold would yield "gold cubic pyrites" and "green carbonated copper."[63] Thus, if gold and silver mining produced copper debris, copper mining was a much cheaper economic venture than financing the whole process of extracting copper ores.

Nevertheless there were also some mines and mining enterprises that concentrated solely on the extraction and production of copper. Some were responding to price increases on the international market. When the price of copper went up entrepreneurs in Peru became interested in processing or transporting the copper debris to ports and marketplaces. This fact is reflected in figure 2.4, which shows sharp

TABLE 2.4
Price Indexes for Copper, Silver, and Gold, c. 1870

Mineral	Price Index	Rate (pesos per quintal)
Copper Ore	66	1
Copper	150	2.2
Silver	43,400	657.5
Gold	675,700	10,237.8

Source: Hunt, *Price and Quantum Estimates*, 63, table 23.

45

fluctuations of copper exports from year to year. Silver mining in particular generated a large amount of copper debris, which sometimes was thrown into the market when international copper prices were up; copper in all forms was kept in storage when prices were low. According to Carlos P. Jiménez, this would explain the copper export boom of the 1850s and 1860s (also reflected in Bonilla's statistics).[64]

> For the past seven years [since 1853] the price of copper has reached and maintained an extraordinary £125 per ton, and this stimulated the export of some raw ore rich in copper; however, the transportation costs were so high that only a small profit margin existed, and as prices later decreased the attempts to maintain this business were abandoned. These attempts, however, encouraged the exploration of ore deposits, so that when copper prices went up again in 1872 from £74 to £100, the exploitation of these deposits was renewed with greater enthusiasm, resulting in the annual exportation of two to three thousand tons of ore with a 20 to 30 percent yield—until the Chilean occupation occurred and this business stopped completely.[65]

Prices for refined copper were even higher, so we would expect a greater production of and demand for it. Refined copper was also the metal in great demand at the Lima mint, which used copper to facilitate the coinage of gold and silver. So, where was this copper mined and produced?

I have already mentioned the mixed production of silver and copper in Cerro de Pasco. There were other production centers, however, from which, rather than being dependent on silver, copper alone was extracted. Profits, of course, were lower in these mines, which formed a less important part of the copper-mining economy. These were areas, furthermore, in which the high quality of the copper ores, proximity to ports and other cities, and availabilty of manpower, transport, and capital permitted the exclusive production of the red metal. The most remarkable of these cases was that of the Ica region, on the southern Peruvian coast.

According to Carlos Jiménez, during the copper boom of the 1850s "most of the exported ore came from the copper-producing areas of Ica and Nazca."[66] During the same years in the area of Cauza, as in Ica, the Desengaño mine was in full operation. This mine "had been worked before by different entrepreneurs [industriosos] on a small scale, reaching a depth of 20 meters and the length of 30 to 40 meters, closely following the copper veins."[67] In 1878 this same copper mine, now called Peru, was still operating at full capacity, creating a whole economic network within which the mining station of Ica was also built and in continuous operation. The Peru mine had attracted new capital investments, mostly Chilean. It featured modern transport (rail cars pulled by animals, good new roads) and employed Chilean, Italian, Chinese, and Peruvian workers.[68] The following year Alejandro and Geraldo Garland formed a "commercial collective society" (sociedad colectiva comercial) with 200,000 soles in capital, 25 percent of which consisted of the mining center of Cauza, which was going to be worked by the company.[69]

In the first half of the nineteenth century, other entrepreneurs pursued the exclusive extraction and production of copper. The examples of the Juan Francisco Izcue enterprise and, later, of the Pflucker family in Morococha illustrate the difficulties facing such ventures.

Juan Francisco Izcue was a merchant from Lima who first invested in the production and exportation of cotton and wool. Later he shifted his attention to the copper ores that, according to Carlos Pflucker, "abound in this country." According to his own recollections, however, he faced several problems: the long distance between the mining districts and the coast, a lack of manpower, a dearth of subsistence goods, high wages, scarcity of and high prices for transportation, and interruptions in the traffic caused by the political oscillations in the country.[70] Finally, with the help of the Pflucker family and especially of Carlos Renardo, he formed the Compañía Peruana en Minas de Cobre. The company sold raw and calcinated ores (minerales crudos y calcinados), although it had not solved some of the previously mentioned problems, so its operating expenses were nearly equal to its sales.

One intractable problem was transportation by muleteers. The company had to pay the muleteers that covered the route Matucana-San Mateo in advance. However, in the summertime the muleteers did not even accept these payments because they earned more money transporting ice from the Andes to Lima. The company thus hired a muleteer team from Piura, for 10,000 pesos, which reduced its available capital for the exploiting of copper.

Another problem was the lack of manpower, or "arms," as manual labor was called then. The company had to recruit peasant workers from Jauja, who used a system called *enganche* (hooking) that required the collaboration of the local authorities, especially governors, to behave as guarantors *(fiadores)*. Furthermore, in the view of Carlos Renardo Pflucker wages were too high: 4 reales per day per *apire* (porter), 5 reales per day per *barretero* (digger who works with an iron bar, with a *barreta*). To decrease their monetary costs these mineowners paid the workers' wages partially in kind, in subsistence or consumer goods, valued above their regular prices. This was, according to Pflucker's own testimony, an old custom proper to mining mills or *"haciendas minerales."*[71]

To obtain more workers without turning to the local enganche system, the Pfluckers spent 4,000 pesos to bring ten men from Germany. They also hoped this would stimulate European immigration to Peru. But, after the hard work and mistreatment they experienced in the mines, the German workers did not last long. In the end they sued their contractors.[72]

The Pflucker family nevertheless persisted in the mining business. In 1878, for example, from the 114 mines working in the Yauli area, 25 belonged to Carlos M. Pflucker and Brothers, although most were silver mines.[73] The Pfluckers had already abandoned the idea of producing exclusively copper and were instead yielding to the prevailing practice of the era: producing mainly silver and copper only as a by-product. With this new business logic the Pfluckers remained a powerful mining family in the area, and would continue to do so until the next century.

I have found similar evidence of copper production in the northern region of Cajamarca. In 1833, for example, the local mining deputation reported that José Manuel Cavada, one of the most powerful businessmen in the area, worked copper mines.[74] And Mariano de Rivero reported that in the Andes between Arequipa and Puno (particularly in the provinces of Lampa, Chucuito, Huancané, Azángaro, and Huacullani, all located in the department of Puno), were copper mines and copper beds, although very few were in full production.[75]

Copper, then, was another of the significant mining sectors in the nineteenth-century Peruvian economy. Its low prices and its low commercial value, however, made it less important than silver and gold. Copper, then, was less in demand, less sought after by prospectors, and less attractive as a business venture for Peruvian investors. But copper was widely available, and its cost of extraction could be absorbed by the costs of silver mining. In any case, large amounts of copper—around 4,000 tons of ore and 2,000 tons of refined copper per year—were exported throughout the century to different destinations, including Great Britain, France, the United States, and Germany. There was also some domestic consumption, which articulated mining production with craft and blacksmith work.

Let us turn now to the last of the predominant mining goods in the nineteenth-century Peruvian economy: tin. Like copper, tin ore was extracted from the bowels of the earth and processed into metal. Processed tin was white, like silver, although tin was not a precious metal but, like copper, an industrial one. Tin served, and still serves, as a raw material for the overseas manufacture of industrial goods, particularly cans to preserve a wide variety of goods.

Tin was another significant part of the mining sector in nineteenth-century Peru. I have found evidence that Peru was a tin exporter since late colonial times. Before 1815, for example, tin was exported without duty from Peruvian ports, but since that year, according to the *Arbitrio temporal*, an export duty of one peso per quintal was levied.[76] Like copper, tin was among those goods that started to gain commercial significance during the Bourbon reforms—the colonial revitalization of the Spanish Empire after the mid-eighteenth century—a social process certainly linked to the commercial and industrial revolution taking place in England and other parts of northern Europe.[77] Nevertheless we have information on its exportation, as well as the markets to which it was exported, only after Independence (since 1825).

As with those who have chronicled the history of other mineral industries, the two authors who describe Peruvian tin exports, Shane Hunt and Heraclio Bonilla, are both interested in following the general cycle of Peruvian exports. In his study of the economic relations between Great Britain and Peru, Bonilla offers figures obtained from tables of British imports from Latin America in the nineteenth century.[78] Hunt extends this vision by considering shipments of Peruvian tin to France, the United States, Germany, and Chile.[79] Exports to Chile, however, must be treated with caution because, as the port of exit for Peruvian exports bound for the North Atlantic basin via Cape Horn, Valparaiso was a port frequently used by Peruvian foreign trade during the nineteenth century. As usual, Hunt's figures are higher than Bonilla's. Finally, I have found in my own research data on Peruvian exports of tin to the United Kingdom for the 1830s and 1840s, which were probably included in Bonilla's work.[80] Figure 2.5 has been constructed from all these sources:

As is obvious from the graph, Peruvian tin exports during the nineteenth century then were almost nonexistent. Until the end of the century tin exports never surpassed 1,000 tons per year, although the export of tin ore does gain significance from the 1850s on. However, tin exports experienced a setback after 1894, perhaps due to competition from massive exports of Bolivian tin.

The paucity of tin production and exports during the nineteenth century seem even more anemic if we compare them with those of copper presented in figure 2.4. For six different years in the 1830s and 1840s, no tin ore whatsoever was exported; no refined tin for fourteen years between the 1830s and the turn of the century. Hence tin, like other industrial minerals—lead, coal, and iron—contributed little to the mining economy of nineteenth-century Peru.

The Peruvian mining economy then, apart from the precious metals (gold and silver) and to some extent copper, did not involve the production of other industrial minerals that, paradoxically, were beginning to be highly appreciated in the international market. Moreover, countries such as Great Britain, Belgium, France, Germany, and later the United States would use industrial minerals such as coal, iron, tin, and copper on a massive scale to produce steel, machinery, or other modern industrial products.[81] Thus, about tin and other minerals, it is fair to say that nineteenth-century Peruvian mining history is the history of lost opportunities, or perhaps the history of the inability to exploit important natural resources whose existence was well known in the country by those who could have facilitated such exploitation.

Tin prices, as one might expect, were also very low. More so if we compare them with those of silver and gold. On the international market, tin ore and refined tin were valued at 0.04 pesos and 0.11 pesos per marc in 1870, compared to 8 pesos per marc for silver, and 130 pesos for gold.[82] These low prices contributed to the lack of commercial interest for tin within the Peruvian economy. And only a very favorable increase in the price of copper, in 1861 and 1862, made that metal attractive and competitive with silver. During those years copper represented 15 percent of all Peruvian exports to Great Britain and France, compared to 11 percent for silver.[83] But this was an exceptional moment, and nothing like it happened with tin. Its history, then, is that of an ephemeral mining good trying to survive as a self-sufficient economic sector during the whole century.

Peru's small tin exports nevertheless had markets. If we subtract from Hunt's data those of Bonilla, we learn that 42 percent of Peruvian tin was exported to Great Britain during the nineteenth century,

FIGURE 2.5

Peruvian Exports of Tin Ore and Refined Tin, 1830–1900

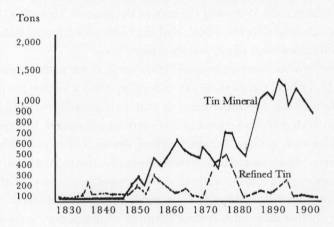

Sources: Hunt, *Price and Quantum Estimates*, 38–40, table 14; Bonilla, *Mecanismos de control*, 193, table 20; and "Foreign and Colonial Merchandize Imported into the United Kingdom from Peru," in *Accounts and Papers* of the British Parliamentary Papers 39 (1849): 372, PRO, FO.

while 58 percent went to France, the United States, Germany, and Chile. The irregularity of these deliveries, however, is notable. They vary from year to year, from hundreds or even thousands of tons to zero. The United States and France were more dependable markets, whereas the British market—the largest importer of Peruvian tin— was the most volatile. The volatility of tin exports reflects the dependent nature of tin mining. Deposits of tin ore were exploited mostly in conjunction with silver mining, and even then tin was put on the market only occasionally, following irregular cycles, depending on prices. Tin production, thus, did not follow its own dynamics.

I also suspect that not all the tin exported from Peru was Peruvian, but was in part Bolivian. It is impossible to know the proportion of tin exported from Peru that originated in Bolivia, but it is clear that Bolivia was, and still is, a great tin-producing country. The Peruvian mining tradition historically has been based on the existence and extraction of silver and copper deposits. The Bolivian mining tradition has been based on the production of silver and tin. In fact, tin was, from the last part of the nineteenth century, Bolivia's main export good.[84] Thus, whereas the focus of Peruvian mining in the last two centuries passed from silver to copper, Bolivia passed from silver to tin.

These historical processes, of course, have to do with an obvious geological reality. In Peru the mining deposits of silver contained large amounts of copper (as in Cerro de Pasco), while in Bolivia silver is associated with tin (as in the mining areas of Llallagua, Uncía, Aullagas, Potosí, Porco, Huanchaca, Portugalete). Thus, the crisis of silver mining in both countries, and their entry at the end of the last century into the world capitalist market of industrial goods—after the so-called second Industrial Revolution—gave Peru and Bolivia the opportunity to exploit their abundant mineral deposits of copper and tin, respectively, goods that in the past had been extracted as by-products of silver production. If the history of Bolivian mining since the end of the nineteenth century is the history of tin mining, and the Peruvian history that of copper, does it not follow that the above statistics include Bolivian tin? Tin exported from Peruvian ports but really extracted from Bolivian mines?

Heraclio Bonilla has pointed out that British commercial records designate as Peruvian those exports that arrived in ships under the Peruvian flag, as well as those goods that came from the port of Arica, without discriminating their real origin. Before the development of the port of Cobija in the 1840s, and for some time after, Arica was, apart from a Peruvian port, the exit to the Pacific Ocean for Bolivian exports (while Buenos Aires was the exit port to the Atlantic).[85] Only with Cobija in active operation was it possible to differentiate clearly between Peruvian and Bolivian exports. Thus, "Peruvian exports included many goods introduced from the interior of Bolivia, particularly tree bark and tin."[86]

If this is true, a portion of the tin export figures shown above includes Bolivian tin that was traded in southern Peru and exported through the port of Arica. The work in Bolivian mines was complemented by Peruvian or foreign traders who bought the mineral and packed it for export in southern Peruvian trading cities and ports. Typical of these traders was Francisco Sales Vidal, a Peruvian merchant who, in 1821, immediately after Peru declared independence from Spain, demanded authorization to transfer 130 quintals of tin from the pailebot *Dos amigos* to the brigantine *Columbia*, in the port of Callao. Sales Vidal, who later developed trading networks with Cerro de Pasco, had purchased the tin in Arica, from where the *Dos amigos* was coming. The *Columbia* needed to make the transfer in Callao because it was going to sail later for Calcutta, where the tin would be sold.[87] In this way the regional and intercoastal trade (between Bolivia, the port of Arica, and Callao) was articulated with the great transoceanic trade (between Lima and Calcutta). Mining products in Bolivia were traded in southern Peru, making the recent border created by the wars of Independence an artificial division. The southern Andes of Peru and Bolivia remained an economic and social unit, and tin production and trade was one of its components.[88]

Typically, tin exports came not in a series of small deliveries but in a few large ones. The yearly average of tin exports to the three largest international markets in the 1830s (Great Britain, France, and the United States) was 161,709 marcs. In just one trip Sales Vidal transported 26,312 marcs of tin, 66 percent of all the tin exports to Great Britain in 1832. This characteristic reflects tin's status as a secondary product, dependent on silver extraction. When large tin deposits accumulated in any mining area a large delivery was made (assuming that the price at that moment was warranted the effort).[89] If prices were low, tin could simply lie as debris, awaiting better prices. This wait could last one year, even several, as borne out in the export statistics. Nevertheless, at least a small amount of tin had to be produced since tin, like lead, was necessary for amalgamation, a vital step in the process of silver refining.[90]

Twentieth-century records show that Peruvian tin was produced in Cerro de Pasco, the highlands of Lima (Huarochirí), Carabaya, and Sandia in Puno. Except for the Puno mines, which mostly produced tin and gold, those of Cerro de Pasco and Lima contained drosses of lead that in 1936 yielded 99 tons (435,604.4 marcs) of tin.[91] Minor amounts of tin, again, were found in other ores, such as lead ore, that also contained some copper and silver. Lead was another of the minerals that were present in Peruvian mines but remained almost unexploited in the nineteenth century. The mineowners of Cerro de Pasco worked and looked mostly for silver, but in exploiting it they would have obtained ores that also contained (and even in larger amounts than silver) copper, tin, or lead. Sometimes these ores were exploited commercially and exported to various markets. That was the case of copper and, secondarily, of tin. Copper was also more abundant and commanded better international prices than tin or lead. Peruvian mining engineer Mario Samamé Boggio has written that "the existence of ore deposits of tin in the Peruvian central region, that is, in Cerro de Pasco and Tambillo, is more sporadic," while copper in the same region "has more deposits of complex minerals."[92]

Although tin represented only a small part of the mining economy in nineteenth-century Peru, it was nevertheless important for daily domestic life, since it could be combined with copper to make bronze. The precolonial and colonial history of Peru has innumerable examples of this use of tin.[93] There is no reason to suppose that this traditional use of tin disappeared in the nineteenth century, although the preponderance of gold and silver production probably obscured it. Indeed, tin, though not a highly dynamic export sector, maintained a constant and steady presence at the national and local levels throughout the century.

Mining and the Nineteenth-Century Peruvian Economy

The four mining sectors that we have explored here—silver, gold, copper, and tin—were in active production during the nineteenth century, although—with the exception of silver—none of these industries was extremely dynamic. As we have already seen, it is difficult to estimate a specific commercial value for each mining product since throughout the century each has experienced cycles of economic growth and decline, as well as fluctuations in price. Silver had four cycles of growth and crisis, in declining order of magnitude, from 1771 to 1900. Gold had two periods of growth and crisis until 1850, and later a declining trend that was to change only at the end of the century. For copper and tin, the industrial metals, the second half of the century showed growing export trends—an indication of the economic transformation and technological change that was occurring at the end of the century in Peruvian mining, where the industrial metals, particularly copper, were gaining ground over the precious ones, particularly silver.

We will gain a clearer picture of nineteenth-century Peruvian mining if we estimate an average yearly value of the economic importance of these four mining sectors. These estimations, however, are only fictitious calculations. In reality, these four mining sectors fluctuated yearly in their values, prices, and volumes of production and exportation. These calculations, nevertheless, illustrate orders of magnitude for the commercial relevance of these four mining sectors.

Silver, with an average production of some 350,000 marcs per year and prices around 10 pesos per marc (before the 1870s crisis), can be valued at more than 3 million pesos per year on average in the nineteenth century.[94] Silver was certainly the queen of the Andean mining economy; it was four times more valuable than its most competitive sector, copper. Gold production averaged a little less than half million pesos per year. Copper exports, with an average annual value of 608,000 pesos for refined copper and 133,000 pesos for copper ore, were worth some 741,000 pesos per year. Finally, tin exports were worth 253,000 pesos per year (121,000 pesos for refined tin and 132,000 pesos for tin ore).[95]

Unfortunately, my data are less than perfect and not fully comparable across mineral products and mining sectors: as we have seen before, silver data are based on production records in Cerro de Pasco and estimations of the national figures, gold on production and minting records, while the figures for copper and tin are based on exports. Production, minting, exports: the three are different dimensions of one economic reality, that of mining in nineteenth-century Peru. Further research is required to fill the gaps.

But if by the measure of value silver was the largest mining sector, by volume of production copper dominated. The 741,000 pesos per year of the copper economy represented the exportation of some 4,000 tons of copper ore and 2,000 tons of refined copper—a total of 26 million marcs, compared to 350,000 marcs per year for silver, although refined silver or, in other words, silver metal. Obviously the amount of silver ore extracted was larger, and it contained a great proportion of copper-bearing minerals. In any case the dimensions and the magnitude of both economies in terms of mass, in terms of volume, are now inverted. In volume the copper industry was larger, but not in value. A raw good, copper, was extracted and exported in large quantities, compared to a refined precious metal, silver, that required a previous transformation process and had, then, a larger value added. However, in the production process the extraction of silver ores was the main economic goal behind both mining sectors, silver and copper, because copper was also extracted as a by-product of silver production. Silver was indeed the leading force in nineteenth-century Peruvian mining.

This last observation leads us to reflect in more detail on the social conditions surrounding the production of silver, gold, copper, and tin.[96] Certainly silver and gold were high-value commodities produced by small-scale industries, while copper and tin were produced on a large scale, using, since the second half of the nineteenth century, modern transportation, the railroad. But how important was mining compared to the guano industry, which according to many authors produced the leading export commodity in nineteenth-century Peru?

Silver and Guano, Exports and Domestic Development: Which One Leads the Train?

I have described the various mining industries that developed, with greater or lesser success, in nineteenth-century Peru. As a whole, mining was not the "annihilated industry" that Carlos P. Jiménez wrote of, nor were the mines closed, as Denis Sulmont claimed. On the contrary, mining was a lively economic activity with distinct sectors, each with its own rhythm of evolution, its own particular dynamism. Certainly, nineteenth-century Peruvian mining was in some sense impoverished, particularly if we exclude the precious-metals sector. And more so if we compare it with mining developments in Europe or the United States, where a technological and productive revolution, the so-called second Industrial Revolution was taking place. In Peru there was nothing similar to the rapid growth of the coal, iron, and steel mines and mills that were revolutionizing the capital-equipment sector and, in general, the heavy industry sector in Great Britain, Belgium, France, Germany, and the United States. Nor did Peru fully utilize industrial metals such as copper, tin, zinc, and lead. In Peru these developments had to wait for the twentieth century.[97]

Nevertheless, mining was one of the leading productive sectors of the Peruvian national economy in the last century; and silver mining was competitive with other dynamic industries such as guano, nitrates, or agricultural exports like sugar or cotton. Guano was the center of a commercial explosion in Peru beginning in the late 1840s and, accordingly, has been the focus of a large volume of historical literature in the nineteenth and twentieth centuries. The lower estimates suggest that between 1841 and 1878 total exports of guano were worth some $600 million (Levin) or 648 million soles (Rodríguez).[98] The higher total estimates range from 750 million pesos (Hunt) or 763 million pesos (Bonilla) to 814 million soles (Tantalean Arbulú).[99] I will assume, then, an average of around 700 million pesos or soles.[100]

Silver production (see fig. 2.1) reached a total value of £42,986,004 between 1833 and 1898, or some 380 million pesos, considering the variations in the exchange rate. These variations must be considered because the cycle of silver was longer than that of guano, which closed early, in the late 1870s, while silver production continued steadily until the next century. Therefore, the British pound sterling was worth more after the crisis of the 1870s,[101] a crisis that was also the result of the economic, financial, and political breakdown of the guano industry. When we divide 380 million pesos by £42 million, we get an exchange rate of 9 pesos per pound sterling. This rate is probably too high, in which case silver production between 1833 and 1898 was worth less in terms of Peruvian soles. But in any case the dominance of guano is clear. Economically, guano was almost twice as important as silver in terms of raw export revenues. But the commercial boom generated by guano was concentrated in a mere thirty-seven years. Silver production, on the other hand, was relevant for the national economy even before 1771. And it continued throughout the nineteenth and the twentieth centuries, uninterrupted by dramatic breakdown, except for a short period during the wars of Independence. Guano, on the contrary, disappeared from the economic scene and was replaced after its boom by a new extractive industry, nitrates. A methodological question, then, should be raised here: How, and especially when, should the silver cycle be calculated so that it may be compared with the guano cycle? The figures presented above (700 million pesos for guano against 380 million for silver) artificially make the comparison between 1841–1878 for guano and 1833–1898 for silver, favoring, of course, the guano revenues, because in other moments during the nineteenth century (1800–1840, for example, or 1881–1900) the impact and the revenues of guano were simply zero or almost zero.[102] The point is that silver had a minor impact during the guano years but played a longer role in the development of the national economy throughout the century. Figure 2.6 shows the evolution of the guano and silver economic cycles, and their contribution to the national economy.

Furthermore guano was an economic activity concentrated on the Peruvian coast or on off-shore islands, the *islas guaneras;* its impact therefore was strongest in Lima. The silver-mining economy, on the contrary, was a productive activity based in the interior of the country, in the Andean highlands, where most of the mining centers, and especially Cerro de Pasco, were located.[103] The production of silver, therefore, demanded the organization of an economic network, an internal market, that linked different areas in the interior of the country.

Silver, however, was only one component, although the most powerful one, of the mining economy in the nineteenth century. Gold was also produced in significant amounts. Copper and tin were produced to be exported on a small scale, and mercury mostly for domestic consumption (particularly in silver mining). And other mining production (coal, lead, iron) also existed, although its history is more one of the misuse of natural resources than of their productive use— paradoxically, at a time when the international economy was moving very fast toward the large-scale use of these goods to produce capital equipment and machinery. Why then was the Peruvian mining economy still based on the production of precious metals while the international economy was moving toward the efficient use of industrial metals, especially iron and coal, to produce steel?

This question raises the issue of the backwardness of the whole Peruvian economy in comparison with developed countries, which I will not attempt to answer here. Rather I will focus on mining. In the mining sector Peru continued to produce precious metals (gold and silver) throughout the century, maintaining a productive structure whose origins were in colonial times. At the same time new industrial metals such as copper and tin were only beginning to be significant. Some of these new industrial metals, such as iron, would not be produced on a significant scale until the twentieth century. In this sense Peruvian mining was still behind some mining industries in neighboring countries, such as Chile, where since the 1830s copper was the main mining export, or Bolivia, where after a transitional period be-

FIGURE 2.6

Silver and Guano Export Cycles in Nineteenth-Century Peru
(current pesos or soles, in thousands)

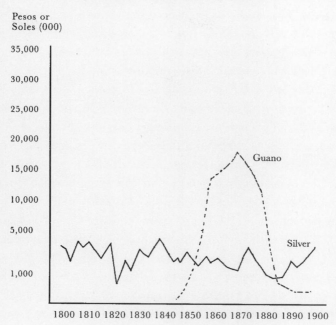

tween 1850 and 1873, tin acquired the leading role in mining production and exports.[104]

In Peru the predominance of copper over silver mining would only
occur at the turn of the century. The extraction and industrial refining
of copper started to be competitive with silver after the 1890s. In 1897
copper displaced silver as the main mining output of Cerro de Pasco,
the most productive mining center in the country. And in 1906 the
value of copper production nationally reached £996,055, while that of
silver was £972,958. This was the first time in Peruvian history that
the value of copper production was higher than silver at the national
level.[105] From then on, the copper industry would keep its predominance. This fact however was the result of an economic process that

started with the building of railroads and the establishment of industrial enterprises in Casapalca, Morococha, and later Cerro de Pasco, beginning in the last third of the nineteenth century.

Thus, I hope to have convinced the reader that throughout the nineteenth century mining was a moderately important sector in the Peruvian economy, and that during most of the century the value of precious metals exceeded that of industrial metals. That was not the case at the end of the century, an era of transition to a more industrialized mining economy in which the arrival of foreign capital played a crucial role. The crisis of silver mining and the beginning of a copper boom would alter the relationship that existed traditionally between these two sectors of the mining economy. Table 2.5 illustrates this quantitative and qualitative transition from silver to copper, from precious to industrial metals in the early twentieth century.

As we can see, copper production grew from 9 million kilograms in 1903 to 27 million in 1910, the latter with a value of more than 15 million soles. This was a three-fold growth in production volumes, and nearly a four-fold growth in production values. And, although the production of silver grew too in those years—from 170,000 to 258,000 kilograms—the rate and dimension of this growth was less. Thus, by 1910 the value of silver production was barely half that of copper. A new century was awakening. The industrial mining era of this Andean country had commenced.

TABLE 2.5
Volume, Price, and Value of Copper, Silver, and Gold Production in Peru, 1903–1910

Years	Copper			Silver			Gold		
	Volume (kilog.)	Prices (soles per kg.)	Value (soles)	Volume (kilog.)	Prices (soles per kg.)	Value (soles)	Volume (kilog.)	Prices (soles per kg.)	Value (soles)
1903	9,497,000	0.50	4,768,240	170,804	33.95	5,799,630	1,078	1,346.98	1,452,050
1904	9,504,000	0.53	5,046,040	145,166	36.57	5,308,750	601	1,249.61	751,020
1905	12,213,000	0.59	7,259,010	191,476	37.62	7,204,440	776	1,379.66	1,070,620
1906	13,474,000	0.79	9,960,550	230,294	42.24	9,729,580	1,247	1,365.95	1,703,350
1907	20,484,000	0.78	16,116,720	206,546	42.08	8,692,280	777	1,366.85	1,062,050
1908	19,854,000	0.51	10,236,310	198,888	32.74	6,511,610	997	1,355.42	1,351,360
1909	20,068,000	0.54	10,839,920	209,656	30.50	6,396,560	554	1,364.67	756,030
1910	27,374,000	0.55	15,141,240	258,565	30.76	7,953,709	707	1,366.50	966,120

Sources: Pablo Macera, *Estadísticas históricas del Perú: Sector minero (volumen y valor)* (Lima: Centro Peruano de Historia Económica, 1972), 20–28, tables II.1.1, II.1.2, II.1.3, II.1.5, II.1.7, II.1.9. His sources are *Boletín del Cuerpo de Ingenieros de Minas del Perú, Anuarios de la Industria Minera en el Perú, Anuario Estadístico del Perú, Boletines Oficiales de Minas y Petroleo,* and *Extracto Estadístico del Perú.*

Chapter 3

Mines, Mineowners, and Mine Workers

Mining as Social System and as Market

AS SHOWN IN the previous chapter the mines of nineteenth-century Peru produced a significant and steady output—primarily silver, gold, copper, and tin. From the quantitative dimension of economic history, let us now turn to the social realm, to the mines and the working conditions therein, as well as to the lives of both the people who labored in them and who owned them. But first, a simple question: What exactly was a mine in nineteenth-century Peru?

The answer varies depending on whether one views the mine as a productive phenomenon or a legal entity. A mine obviously was a workplace in which people extracted raw materials or, in another definition, natural goods, from the depths of the earth: mineral ores or rocks that were later transformed into metals such as silver, gold, copper, or tin. According to the Nuevas Ordenanzas de Minería of 1786—the colonial legal framework that ruled the mining industry for three-quarters of the nineteenth century (until 1877)—a registered mine had to be at least 10 varas (8 meters) deep and 1.5 varas (1.25 meters) in diameter.[1] Mineowners, however, usually extended these dimensions in their drive for minerals. Large mines with tunnels

MINES, MINEOWNERS, AND MINE WORKERS

branching off at several meters of depth were called *socavones*. But miners frequently did not find much ore after digging more than eight meters down. These small, superficial mines were usually called *bocaminas*.

The national congress approved the Ley de Reforma de la Minería in 1877, but this act did not change the legal organization of the mines completely. The real change occurred in 1900, when the liberal Código de Minería was promulgated, opening mining production to foreign investment and paving the way for modern capitalist enterprises. Thus the minimum dimensions required for a mine remained unchanged until the beginning of the twentieth century.[2]

From this overview of the legal framework of nineteenth-century Peruvian mining we can conclude that liberalization occurred very slowly during the century, and that until the twentieth century mining took place within a colonial legal framework.[3]

Finally, a mine was, of course, a center of production, a workplace that brought together mineowners (or their representatives) and mine workers, the ones who actually extracted from the depths of the earth the minerals whose production statistics were presented in the previous chapter.

The Number of Mines

The data I have gathered about the number of mines in Peru and the kind of minerals extracted from them cover the whole century, from 1790 to 1887. Unfortunately they are not continuous statistics but a set of documents—*censos, matrículas,* and *padrones*—written principally during two periods, between 1790 and 1799 (including particularly detailed documents in 1791), and from 1878 to 1887 (including similarly elaborated documents in 1879). The first moment relates to the Bourbon reforms at the end of the eighteenth century, and the first boom period in mining production in Lower Peru, particularly in Hualgayoc and Cerro de Pasco. The second represents a moment of mining reorganization with the creation of the School of Mining and the promulgation of the new Ley de Reforma de la Minería, in the 1870s, a moment also related to the guano boom. This second moment, however, was interrupted by the War of the Pacific between 1879 and 1884. An uninterrupted series of mining statistics appears only at the end of the nineteenth century.[4]

Colonial statistics are different from republican statistics, so I have constructed table 3.1 only from the former.

The data for 1790 were collected by the Diputaciones Territoriales de Minería, the local bodies of the colonial authority concerned with mining production. The data for 1791 come from reports of the intendants (provincial authorities named by the central government), and those for 1799 from an *estado general* carried out by the Lima Audiencia for the Indies Council in Seville.[5] These documents, then, are quite different from each other, as indicated by the data. The first, being more closely related to the mining sector, is a more reliable source.

Whatever the accuracy of these numbers, at least they give us a certain order of magnitude. They also suggest the kind of mining production that existed at the end of the eighteenth century. The continual oscillation between a mine in operation and its being considered *parada* (paralyzed) was partially due to the nature of the mining work. Peruvian mining at the end of the colonial period, and to some extent in the nineteenth century, was based on small family-owned mines and mine companies that could quickly close down in response to a sharp decline in prices or profits. This fluidity made more difficult an accounting of the number of mines by the territorial deputies or the intendants. We could expect therefore changes year by year, although not usually as dramatic as the sharp increase in the number of silver and gold mines not in operation *paradas* in 1799. These oscillations notwithstanding, the general trend in the data reflects some growth in the number of mines.

Although I have suggested not to trust these numbers completely, they clearly reflect many of the main features of the mining sector at

66

TABLE 3.1
Types of Mines in Peru, 1790–1799

Year	Mineowners	Silver Mines in Operation	Silver Mines Not in Operation	Gold Mines in Operation	Gold Mines Not in Operation	Mercury Mines in Operation	Copper Mines	Lead Mines	Total
1790	706	670	578	63	8	4	n.d.	n.d.	1323
1791	n.d.	784	588	69	29	4	4	12	1490
1799	717	633	1124	55	57	n.d.	n.d.	n.d.	1869

Sources: The 1790 data come from the "Razon de la matricula general de los mineros, minas de plata en labor, minas de plata paradas, minas de oro en labor, minas de oro paradas, minas de azogue en labor, haziendas de beneficiar plata y piruros de beneficiar oro en el Reyno del Peru formado por las particulares recibidas por las Diputaciones Territoriales," 30 April 1790 and published by Fisher ed., *Matrícula de los Mineros*, 2–33. See particularly the "Resumen General," p. 33. The 1791 data come from the references of the British consul in Peru, Charles Milner Ricketts, who mentions that his sources are Intendants' reports (see Bonilla editor, *Gran Bretaña y el Perú*, I, 2); for 1799 the source is the "Estado general de minería de 1799" of the Archivo General de Indias, Seville, published also by Fisher ed., *Matrícula de los Mineros*, 34.

the time. The dominance of silver mining, for example, appears clear. Silver mines (both operative and inoperative) accounted for 94 percent of all mines in 1790, 92 percent in 1791, and 94 percent again in 1799. Even gold mines, the second most frequent kind, easily outnumbered mercury, lead, or copper mines. The mining of precious metals definitely dominated, but the industrial metals were not absent, even in late eighteenth-century Peru. By 1791 there were already mines that produced lead and copper exclusively.

Statistics for the nineteenth century are more varied and complex than for the colonial period. It is difficult to summarize in one table three major documents of that era: the *Estadística de minas* of 1878, the *Minas empadronadas* of 1879, and the *Padrón general de minas* of 1887. In the eighty years that had elapsed since the colonial statistics of the 1790s, scientific methods had been developed and modern field surveys implemented.[6] However, as usual, there is always a margin of error.

Table 3.2 lists mines by type, as given in the *Estadística de las minas de la república del Perú en 1878.*[7]

Although the *Estadística* lists all these mines as operative, we cannot be totally sure of that. Other primary sources indicate that there was always a percentage of mines that were not in operation, what the colonial statistics called *minas paradas.* In 1887, for example, I. C. Bueno's report on Cerro de Pasco revealed that "of 458 mines registered in that *asiento,* 103 were in operation at the beginning of the year but this number has been reduced to 46."[8] In 1878, according to the *Estadística,* there were 674 mines in Pasco, although we cannot be sure that all were actually in operation.

I offer a second methodological observation: the mines in the original sources of the *Estadística* are classified systematically by their type of mineral extracted, but in a more complex way than in the colonial statistics. Nevertheless the *Estadística* still keeps some empirical way of making this classification. In 1833 José Manuel Sorogastúa classified mineral goods according to their appearance *(de*

TABLE 3.2

Types of Mines in Peru, 1878

Type	Number
Silver	933
Coal	167
Copper	144
Petroleum (includes wells)	53
Gold	17
Lead	11
Cinnabar	6
Iron	5
Sulfur	3
Bronze	3
Lime	2
Unspecified	268
Total	1612

Source: Dirección de Estadística, *Estadística de las Minas de la República del Perú en 1878* (Lima: Imprenta del Estado, 1879), esp. pp. 92–93.

acuerdo a su pinta): "llintas, sucos, orangish, syrupy ones, blacks, grays, purples, mulattoes, *almagrados.*"[9] In 1878 this local empiricism was still present in several places, although in others, particularly in Cerro de Pasco, technicians and engineers were already starting to give advice and work alongside mineowners and mine workers, particularly because of the recent founding of the School of Engineers in Lima.[10]

The classification of the 1878 *Estadística* describes minerals as *pacos* (reddish brown), *paco y azul* (reddish brown and blue), *pavonados* (quicksilver), and so on—all local, traditional names to designate ore forms of silver.[11] Thus, my organization of the information in the preceding table includes among the silver mines all those that are mentioned in the document as *galena* (the principal ore from which silver and lead are extracted), *galena argentífera* (silver-bearing), *galena, cal y pirita* (lime and pyrites), *acerillo, galena y tameniana* , *pacos, paco y azul, pavonados , paco y pavonado, plata* (silver), and so on. What I have classified as copper mines include mines of *cobre* (copper), *cobre gris* (gray copper), *cobre y pirita* (copper and pyrites), *cobre y galena,* and so on. I have thus reduced the thirty-eight categories in the 1878 document to twelve, according to the predominance of certain minerals (e.g., silver, coal, copper). In the case of a mine that produces two minerals, say copper and lead or gold and copper, I have assigned to each mineral the value of half a mine. At the end I have added all these values to obtain the results that appear in table 3.2.

Table 3.2 shows again the overwhelming predominance of silver mining. Still, the relative importance of silver mining had decreased since late colonial times. In 1878, for example, only 57.8 percent of all Peruvian mines produced silver compared to 93 percent during colonial times. There was also an increasing number of coal mines, because, among other reasons, of the recent construction of a railroad system in the country. The number of copper mines had also grown, but there was only a scattering of other mines. (Petroleum was being extracted in increasing quantities, but I do not analyze it because it is not a metal.)[12]

The next year, as shown in table 3.3, the number of mines increased significantly. The historical source for this table, however, is different from the source for table 3.2. The document considers registered mines *(minas empadronadas)* and gives consolidated statistics instead of disaggregated numbers.

The addition of nearly 500 new mines from 1878 to 1879 seems too high. There was certainly some growth in the number of mines because these were the years of the guano boom, of price inflation, of the growth of the money supply (particularly in Lima), of the creation and development of a banking system, and of increasing capital formation. Therefore mineral prospecting and development, particularly of gold and silver, was a good investment at the time.[13] But such a sharp increase in a single year probably reflects instead a change in the technique used in accounting. Perhaps the *minas empadronadas* reported

TABLE 3.3
Types of Mines in Peru, 1879

Type	Number
Silver	1195
Copper	147
Gold	21
Petroleum (includes wells)	20
Mercury	8
Coal and other minerals	780
Total	2171

Source: "Estado de la industria según el empadronamiento de las minas empadronadas," in Emilio Dancuart and J. M. Rodríguez, Anales de la hacienda pública del Perú, 1821–1889 (Lima: Imprenta de La Revista, 1902–1926), 17:95–96.

in the 1879 statistics includes mining claims *(denuncios mineros)* as well as real mines in operation.

Table 3.4 presents mining figures from 1887, after the War of the Pacific had taken place, the mining centers, especially in the central sierra (Cerro de Pasco), had been occupied by an invading army, and a general collapse of the national economy had occurred in which the wealth from guano and nitrates was also lost.[14]

Again, I have consolidated data from widely scattered areas from the *Padrón general de minas,* a very specific and detailed mining census begun at the time. The padrones were created in a new effort to gather more accurate mining statistics. Compared to the other two sources we have studied to this point, Peruvian statisticians were getting better at organizing records and information, a trend reflected also in the new population censuses compiled in this period. Thus the history of mining is also the history of the state and government officials perfecting their tools to comprehend the reality of mining, that economic sphere that somehow escaped their economic designs.

Since the information in this latter source is more accurate and has much more detail, I have preferred this time to present some

TABLE 3.4
Types of Mines in Peru, 1887

Type	Number
Silver	1438
Silver and lead	20
Silver and copper	19
Silver and gold	14
Silver and zinc	3
Silver and mercury	1
Total silver mines	1495
Coal	147
Copper	51
Gold	21
Petroleum (includes wells)	17
Sulfur	10
Mercury	3
Copper and lead	1
Lead	1
Bronze	1
Unspecified	31
Total	1778

Source: Ministerio de Hacienda, *Padrón general de minas de 1887* (Lima: Imprenta del Estado, 1887).

disagregated data, as in the case of the silver mines, because the classification of this mining industry is now much more scientific. There is no reference to *sucos* or *pavonados* ores, as in the Sorogastúa account or the 1878 *Estadística,* but instead to the real mineral and metallic components of silver mines. This greater scientific accuracy also reduced the number of unspecified mines from 268 in 1878 to only 31 in 1887. We have to think too that these Peruvian mining statisticians were part of a learning process about accounting tools that gave more fidelity to their records and that, compared to 1878, when for the first time in the postcolonial period mining statistics were collected, a learning decade had already passed; a decade with a war in the middle, but a decade nevertheless.[15]

The predominance of silver mines is again overwhelming, even increasing in proportional terms. In 1887, 81 percent of all Peruvian mines extracted silver exclusively, and 84 percent produced some silver component. After the War of the Pacific this concentration on silver reflects the relatively higher economic yields of the precious metals versus minerals like copper or coal. Nevertheless, coal mines occupied second place in the 1887 *Padrón* (8 percent of the total), retaining the importance indicated in the other two registers, and the number of gold mines is the same as in 1879. Production of other minerals, however, decreased dramatically. Iron mines are not even mentioned in the 1887 document.

The significance of the fluctuation in the total number of mines between 1878 and 1887 is uncertain, but I am confident that the following trends are revealed: growth in the number of mines from 1878 to 1879, a subsequent crisis due to the War of the Pacific, and a recovery toward 1887 to levels close to those of 1878. Silver mines were always dominant and their numbers grew throughout this period: 933 in 1878; 1,195 in 1879; 1,438 in 1887. Silver became more attractive as an investment because of the economic boom of the 1870s, the crisis of guano, and, later, the national crisis. Silver was surely the best investment available after the War of the Pacific, but while silver rose there was a relative decline in the importance and number of coal and copper mines.

Although the number of silver mines grew from 1878 to 1887, silver output did not, as we saw in chapter 2. Silver production stagnated between 1878 and 1887, and even declined sharply in 1881 and 1882, although it recovered quickly, so that in 1887 its level was the same as in 1878 (see figs. 2.1 and 2.2). The growing number of silver mines in these records, then, foreshadows a growing economic production that would take place soon after, from 1885 to 1898. Until 1887, nevertheless, the growth in the number of silver mines and the stagnation of silver production indicated a decrease in silver production per mine. This is perhaps the first instance of silver productivity slipping in comparison to copper. This first instance of a change in productivity is even clearer if we compare silver with copper outputs during the same years. However, we need better statistics to deal fully with this question. In any case, the falling international prices of silver certainly did not help the recovery of the silver mining industry after the War of the Pacific.

What can we conclude from these data? First of all, as I have shown in the previous chapter, Peruvian mining was a significant and steady economic activity in the nineteenth century. Now the number of mines confirms this assertion. Furthermore, we see a relative and small growth in the number of mines (keeping in mind the inaccuracies contemporary source data are likely to contain) from the late colonial period (1790–99) to the republican (1878–87). On average 1,560 mines existed between 1790 and 1799, compared to 1,853 between 1878 and 1887. This translates to a modest, but by no means remarkable, 19 percent growth in more than ninety years of mining.

Growth in the number of mines, however, is not synonymous with growth in mining production. Sometimes, in fact, an increase in the number of mines could result in a greater dispersion in mining production and therefore a relative decline in production. Thus, the evidence presented above should be integrated with, not placed in opposition to, the quantitative analysis of the previous chapter. We still do not have a full series of tables in mining production, minting, or exportation. Nonetheless, I have presented reliable data on silver production, gold minting and production, copper and tin exports, mercury production, and the other scattered information I could gather on coal, lead, and iron production, a reflection of the importance of all of these metals for nineteenth-century Peruvian mining.

In the entire period, from 1790 to 1887, between one and two thousand mines were in operation in Peru. This number remained stable throughout the century and even increased in the final decades. Silver mines constituted from 92 to 94 percent of all mines between 1790

74

and 1799, and from 55 to 81 percent between 1878 and 1887. These percentages also show a relative diversification of the mining industry in Peru from the 1790s to the 1870s, although silver mining remained the predominant sector throughout the century.

Silver was not only the main product of mining in nineteenth-century Peru, but also the money supply of the nation because much of the silver was finally minted into coins. Instead of looking into how mining production was articulated to the circulation of money in nineteenth-century Peru,[16] I turn now to the people who made a living from the ownership of these centers of production, of these workplaces. I will also discuss the other side of the coin: the people who worked the mines. I will add a human quality to what, until now, have been arid statistics. This exercise will take us from the realms of economics and quantitative analysis to social history.

Mineowners, Mine Workers, and the Population of the Mining Centers

We saw that there were 706 mineowners in 1790 and 717 in 1799.[17] Archival and other sources enlarge our impressions of them for the first half of the nineteenth century. In 1825 in Hualgayoc, in the highlands of the old intendancy of La Libertad, 75 mineowners controlled 939 workers, of whom 539 were *operarios de minas* (mine workers), 344 *operarios de ingenio* (refining-mill workers), and 56 *pallaquiles* (independent workers that also refined ore, but with primitive techniques).[18] That year there were 193 mine workers in Cajamarca,[19] one-fifth of those in Hualgayoc. In Pasco two years later, 1827, 60 mineowners controlled 2,428 workers, two and a half times the number of Hualgayoc, although the mineowners were fewer.[20] In 1833 in Huallanca, a mining center *"en la rivera de Huaylas"* (on the banks of the Huaylas River), 16 mineowners were listed, along with 195 workers.[21]

Compared with the data of the late colonial period, these numbers show a decrease. In 1790 there were 103 mineowners in Pasco, 106 in Hualgayoc, and 27 in Huallanca.[22] By the national period (1825–33) the number of mineowners in these areas had decreased by 38 percent on average.[23] This means that if the total number of mineowners in 1790 was 706, in 1830 we would expect that number to be about 440. In 1878, however, 724 mineowners were counted in the whole country. Therefore, if a certain decrease in the number of mineowners occurred in the early years of the national period, because of the crisis produced by the wars of Independence,[24] a small increase occurred, compared with the colonial numbers, in the second half of the century. This growth was even greater if we compare the numbers of 1878 with that between 1825 and 1833. Thus, during the nineteenth century the group of mineowners—perhaps it should be called a class—almost doubled.

However, the number of mineowners is not a good indicator of the evolution of mining production. Given that the number of mines remained more or less constant during this period, a larger number of mineowners could only imply that more entrepreneurs controlled the same number of mines or even fewer, not that mining as an economic activity was on the rise. Many times in the nineteenth century several entrepreneurs joined forces to put a single mine into production. A smaller number of mineowners also could simply reflect a greater concentration of property, of mine ownership. Whatever the case (and I come back to this issue later), the data presented above show that the number of mineowners decreased immediately after Independence, during which a concentration process of mine ownership took place. The number of owners then increased again in the second half of the nineteenth century (precisely, in 1878), when the number of mineowners was larger than in 1790, 1799, or 1830, according to my sources and estimates. In other words, it was larger than ever.

The relative concentration of mine ownership in the first half of the century, the control of labor in fewer hands, was clearly visible in the mining area of Pasco. In 1790 there were 103 mineowners with 2,470 workers, or a ratio of 23 workers per owner. In 1827, however, the ratio of workers to owners had risen to 40 to 1. In other words, in that year only 60 owners employed virtually the same number of workers, 2,428.[25]

One mineowner, José Lago y Lemus, typifies the increased concentration of ownership in this period. At the beginning of the nineteenth century Lago y Lemus was one of the wealthiest mineowners of Cerro de Pasco, employing close to a hundred mine workers in 1827. He was born in 1782 and by the 1830s was a resident of Lima and at the same time a neighbor of Pasco (vecino de la villa de Pasco). In 1838 he wrote his will, my main source of information about his life, which recognized his ownership of the mineral hacienda (refining mill) of Quiulacocha, with "its small lake and twenty mines, equipment and furniture, house and chapel." He was also the owner of the mining hacienda of Pampa Hermosa in Colquijirca and "its mines," half the Remedios mine, one bocamina called San José and eight "cortes de minas" (mine yards) in Paccha, another bocamina in partnership with the Ijurra family, and a third of the twenty bocaminas of Quiulacocha. Finally, he was also the owner of another bocamina in Ventanilla, and a coal mine close to the village of Rancas.[26] Not only did he hold these immense properties, but his will also notes that the Indians of Tusi, who worked for him, owed him 170 pesos.

Other great mineowners at the beginning of the century were Pedro Abadía, José Apotino Fuster, Miguel de Otero, and Félix Ijurra. By the end of the century the wealthiest owners were José Malpartida, Lagravère and his sons, and Genaro Maghela.[27] Thus a small but wealthy and powerful group, or class, of mineowners existed in nineteenth-century Peru, a group that would play an important role in shaping Peruvian politics and organizing Peruvian society.[28] They were the dominant class in the mining centers, although nationally they had to compete with the larger landed elite, with guano merchants, and with other groups who wanted a say in politics, society, and the economy.[29] The previous discussion of the ratio of mineowners to mine workers in Peruvian mining centers gives us more clues to understand the roles of these two basic classes of the mining industry. But I should add here that owners and workers were not the only people living and working in mining centers. There was also a general population that did other things related, or even unrelated, to mining.

As noted above, there were 60 mineowners in Cerro de Pasco in 1827 and 2,428 workers. In 1828, according to the testimony of Mariano de Rivero, the total population of Cerro de Pasco reached between 5,000 and 6,000—twice the total number of mineowners and workers.[30] We would expect that workers and owners lived with their families, and thus that the total population should reach several times that of the working people living in a mining center. Of course, not every worker or owner had a family, but in any case mining centers were also places of residence—cities and towns like Cerro de Pasco.

If mine workers lived by themselves, this means that other working people, performing other economic activities and providing urban services, lived in those mining centers. In Yauli, for example, a mining center in the highlands of the department of Lima, the population in 1827 was composed of 80 mine workers and 10 mineowners, together with a much larger group of people involved in various activities, creating a population of roughly 800.[31]

For 1883 we have a very precise census of the same mining town. According to the manuscript *Registro cívico del distrito*, in 1883, of a total population of about 1,500 people in Yauli there were 77 mine workers *(jornaleros de minas)* and 8 mineowners, but there were also 68 muleteers *(arrieros)*, 57 herders, 57 breeders *(crianderos)*, 18 traders, 5 employees, 3 masons, 2 tailors, 2 blacksmiths and 1 agriculturalist, 1 engineer, 1 temporary worker *(maquilero)*, 1 silversmith, 1 carpenter, 1 weaver, and 1 servant. Each was the head of a household.[32] If we combine the mine workers and mineowners, the total represents barely 6 percent of the entire population. These numbers show that Yauli was not only a mining center but also a herding town that raised animals for trade: llamas, sheep, mules.

Furthermore, the nonmining majority in Yauli (unlike those in Cerro de Pasco in 1827 and 1828) worked both in other activities related to mining (blacksmiths, silversmiths, engineers) and in economic activities independent of mining (herders *[pastores]*, crianderos, tailors, weavers).

According to the demographic census of 1876, the province of Pasco was comprised of nine districts. Among these nine districts, only the mining town of Cerro de Pasco had a large mining population, 53 percent of the town's total, if we include all those people in the census whose name appears with the word *minero* (2,068 people). But the census also mentions 711 jornaleros (day workers), which probably also includes mine workers. If we add this number to the miners, the mining population reaches 71 percent of the total. These calculations, however, take into account only the working population of Cerro de Pasco; the town also had 5,080 people "without profession." This does not necessarily mean unemployed adults, but may also refer to women and children. If we add the people "without profession" to the total population, then those involved in mining were only 31 percent of the total, if we include the jornaleros, or 23 percent without them.[33]

In the other districts of the province of Cerro de Pasco (Caina, Chacayan, Huayllay, etc.) the agriculturalists *(agricultores)*, weavers *(hilanderos, hilanderas)*, and herders *(pastores)* were the majority of the population. These numbers, then, suggest that the Cerro de Pasco region was one of several regions in which a large center of mine workers and mineowners coexisted with a surrounding area full of peasant villages and haciendas (large agricultural and pastoral estates).

In the district of Cerro de Pasco in 1876 there were nineteen haciendas with a total population of 1,604 people. The largest was the hacienda of Tullurauca, with 221 peons.[34] The census of 1876 places the total number of haciendas in the department of Junín at 319; 169

of them were located in the province of Pasco, 110 in Tarma, 21 in Huancayo, and 20 in Jauja.[35] Peruvian historian Pablo Macera has also calculated the total rural population, and the rural population living in haciendas, according to the 1876 population census. He found that in the province of Pasco the rural population living inside haciendas was larger than in other provinces, such as Tarma, Jauja, and Huancayo (in exactly that order of magnitude).[36]

In the provinces of Pasco and Tarma, then, the haciendas, and therefore the landed economy of large agricultural and pastoral estates, were dominant and larger than land controlled by peasant economies, which were certainly in conflict with them.[37] This contrasts with what was taking place in the Mantaro valley, south of Cerro de Pasco, where the peasant economies and the peasant communities controlled the best land of the valley.[38]

In nineteenth-century Peru, then, the Andean landscape was filled with pastoral and agricultural areas that overwhelmed the mining towns they surrounded. Even in the mining towns, nonmining activities occupied larger numbers of people and were at least as important economically as mining. This was true of the pastoral community of Yauli in 1827 and 1883, and although the mining industry employed many people in the town of Cerro de Pasco itself, the surrounding districts and provinces were, again, predominantly agricultural and pastoral. Thus mineowners and mine workers were a small minority not only at the national level, but even in so-called mining regions.

This phenomenon depended, of course, on the size and importance of the mining center in question. The populations of larger centers, such as Cerro de Pasco and Hualgayoc, were devoted more exclusively to mining as an occupation, while in smaller ones, such as Yauli, Morococha, and Caylloma, the mining industry was secondary to other economic activities.[39]

The 1876 census also sorts the population by race. The population of Cerro de Pasco was composed of 1,533 whites, 2,759 mestizos, and 5,875 Indians.[40] There was a close correspondence between race and occupation. The Indian population, of ancient Andean ancestry, was not only the resident majority in the mining city but also the largest working population, while the mineowners were predominantly white, Peruvian creoles of Spanish and other European origin (with a few mestizos), who felt themselves attached to a European cultural tradition. In nineteenth-century Peru, the opposition between owners and workers was not just economic but also racial and cultural. Differences extended to language: the mineowners spoke Spanish and probably a smattering of foreign languages such as English and French; the Indian mine workers spoke Quechua (probably Quechua Wanka) and some Spanish.[41]

The Mining Population and Demographic Evolution

Using diverse sources, but especially travelers' observations and calculations, the Peruvian historian Carlos Contreras has made the following calculations concerning the population of Cerro de Pasco between 1813 and 1900.[42]

As we can see from the figures in table 3.5, the demographic evolution of Cerro de Pasco, the largest mining center in nineteenth-century Peru, follows very closely its economic evolution, as reflected in its silver production (see table 2.1), despite the existence of a non-mining population. The wealth created by mining was a powerful magnet for laborers and nonlaborers, who came to Cerro de Pasco to work in the mines or in other economic activities the city generated as a market and as an urban center. In periods of mining decline and crisis, the population tended to leave the mining center.

Table 3.5 illustrates the instability of the general population. A contingent of migrants was constantly coming and going from the mining center throughout the century, following the economic fluctuations of the mining industry. Part of this unstable and mobile population were seasonal mine workers, a significant migrant population

TABLE 3.5

Census Data and Population Estimates for Cerro de Pasco, 1813–1900

Year	Population
1813	9,000 to 10,000
1821	7,000 to 8,000
1828	5,000 to 6,000
1840	18,000
1849*	11,500
1855	12,000
1859	12,000
1861	12,000 to 15,000
1876*	8,974
1880	8,000 to 9,000
1892	8,000 to 9,000
1895	7,000
1897	5,000
1900	10,000

*Based on census.

Sources: The table is taken from Carlos Contreras, "Minería y población en los Andes: Cerro de Pasco en el siglo diecinueve" (Lima: Instituto de Estudios Peruanos, research report, September 1984; manuscript), 23. Its sources are: for 1813 El peruano liberal nos. 4 and 7, cited in Dionicio Bernal, La muliza (Lima: Herrera Editores, 1978), 19. For 1821, José I. Arenales, Memoria histórica sobre las operaciones y movimientos de la División Libertadora (Buenos Aires: Cultura Argentina, 1920; original ed., 1832), 31. For 1828, the report of Mariano de Rivero, "Memoria sobre el rico mineral," 186. For 1840, Tschudi, Testimonio del Perú, 257–58. For 1849, the newspaper El peruano, 26 May 1849, quoting a census of that year. For 1855, a report of the Prefectura de Junín. For 1859, the German traveler Karl Scherzer, "Visita al Perú en 1859," in Viajeros alemanes al Perú, ed. Estuardo Núñez editor (Lima: Universidad Nacional Mayor de San Marcos, 1979), 115–16. For 1861, the German traveler Friedrich Gerstacker, Viaje por el Perú (Lima: Biblioteca Nacional, 1973; original ed., 1862), 67. For 1876, the Census of Population. For 1880, 1892, and 1895 the reports on Cerro de Pasco written by Maurice du Chatenet, "Estado actual de la industria minera en el Cerro de Pasco," in Anales de la Escuela de Construcciones Civiles y de Minas (Lima, 1880), 9; Ricardo García Rosell, Informe presentado a la Compañía Nacional Minera de Pasco (Lima: Imprenta de El Comercio, 1892); and Modesto Basadre. For 1897, the report of the British consul in Peru, in Gran Bretaña y el Perú, ed. Bonilla, 1:258. And for 1900, the mining newspaper El minero ilustrado (Cerro de Pasco), several issues.

that came from rural areas to work in the mines during periods when agricultural work was scarce. These seasonal workers sold their labor power on a temporary basis,[43] and their existence explains why, in only twelve years (1828–40) the general population in Cerro de Pasco increased by 12,000 people, whereas in only nine years (1840–49) the mining town lost 6,500 residents. And these dramatic fluctuations occurred in a period of mining expansion (1825–42), and later steady decline (1843 on; see table 2.1).

The Swiss traveler J. J. von Tschudi noted this phenomenon and gave the following description of a mine worker in Cerro de Pasco around 1838–1842:

> There are times when *boyas* [boom periods] occur in several mines at a time. Then the population of the town doubles or triples. The mine workers, all of whom are Indians, are divided into two kinds: those who work without interruption all year in the mines, usually in debt to the mineowners because of the cash advances the latter give them, and who are registered as mine workers and form a stable group; and those who come to Cerro de Pasco attracted only by the mining *boyas*, who are generally called *maquipureros*. They usually come from distant places and come back to their lands when the mines do not yield as much.[44]

Some of the migrant, or temporary, mine workers of Cerro de Pasco came from as far away as the provinces of Pallasca and Pomabamba in the department of Ancash, several hundred kilometers to the north, and Huancavelica in the south (department of Huancavelica). But the largest contingent of Cerro de Pasco mine workers and residents came from nearby (the province of Pasco, the provinces of the Mantaro valley). Between 1845 and 1900, according to parish records of the Pasco area, particularly of the Chaupimarca parish, 19 percent of the migrant residents in Cerro de Pasco came from the province of Pasco, and 34 percent came from the Mantaro valley, south of Cerro de Pasco, from the provinces of Huancayo, Jauja, and Concepción.[45]

The Mantaro valley was the main source of Cerro de Pasco residents and workers for the period studied by Carlos Contreras (1845–1900).

The fact that residents and mine workers came from the old mercury-mining area in Huancavelica (5.9 percent of all migrant residents in Cerro de Pasco), from Huaraz (1.2 percent), Lima (3.7 percent), and, of course, Yauli (1.0 percent) proves that there were strong communication and migratory networks linking several different mining areas. The campesino workers who left their agricultural and pastoral areas and went to work in the mines moved from one mining center to another trying to find work. Thus many of the residents and workers in Cerro de Pasco came from other mining provinces such as Cajatambo, Canta, Pallasca, and Yauli. Many of these migrant, seasonal workers were peasant herders or farmers who traditionally cultivated potatoes or corn, or kept animals such as llamas or sheep, and depended on the work in the mines to complement their agricultural income.

Sometimes the search for mining work was a family-based strategy. Some members of the peasant family cultivated the land in their villages of origin, while others looked for temporary work in the mines. This temporary work occupied only a portion of the year, a few weeks or several months, and often lasted just a few years, but it was not permanent. The transitory peasant-miner never became a true proletarian during the nineteenth century. In 1835, for example, the deputy of mines in Cerro de Pasco complained to the subprefect that there was "an absolute scarcity of workers [operarios]," because they "have gone back to the valleys" *(de haberce [sic] ausentado a las quebradas)*.[46] The temporary worker in Cerro de Pasco, then, was a campesino, a land-based operario, who considered mining just a seasonal job. This characteristic of the mine-working population, of course, was to create tensions between mineowners, interested in a fixed, totally dependent labor force, and mine workers.

This lifestyle of the nineteenth-century mine worker—half miner, half peasant—is an extremely important feature of the mining economy and society and has attracted the attention of several authors.[47] One speaks of a period in which "peasants confronted commerce" (1860–1900), and a following one in which they "confronted industry" (1895–1930).[48] But during the period that the peasants of the Peruvian central sierra "confronted commerce," some were already working in the mines, in a seasonal, temporary form. The previous table reflects this impermanence of the mining labor force, the frequent oscillations of its numbers, the lack of a stable and growing proletariat in the mines.

The demographic evolution of Cerro de Pasco in the nineteenth century, then, is a faithful reflection of the mining economy as a whole. As we have seen in the previous chapter, Peruvian mining never experienced a period of rapid growth in the last century. In fact, on the whole it declined; at least silver did, the main product of the mining industry. The population of mine workers was mostly a temporary and migrating labor force that divided its time between agricultural work in the countryside and mining work in the various mining centers of the nation.[49]

Throughout the nineteenth century, then, the working population for the mines was clearly composed of two segments. One was a stable, specialized group settled permanently in the mining areas, particularly in Cerro de Pasco. This group included barreteros, technicians, and foremen. The largest group, however, went to the mining centers only temporally to earn cash to contribute to their peasant-based economies in Jauja, Tarma, Yanamarca, in the Mantaro valley, or in other faraway places. A true mining proletariat, however, was produced only through a slow, cyclical process and came into existence only at the turn of the century, when the majority of the mining working population clearly settled in Cerro de Pasco and mining became a constant in their lives. It was only then that mine workers started to organize in trade unions, to participate in strikes, and to develop a working-class culture.[50] The impermanence of this slow transition of nineteenth-century mining, during which workers left the mining centers and returned to the countryside to live off the land, is what gave the enganche process its importance. The enganche was an effort at the forced creation of a more stable and settled mining working class. We will focus later on the enganche and the more rapid transition that the labor force experienced at the end of the nineteenth century.

I turn now to an estimate of the total number of mine workers and mineowners in nineteenth-century Peru. In 1799 there were 8,875 mine workers in Peru, according to the estado general, and 717 mineowners.[51] In 1878 the number of workers had decreased to 5,071.[52] In 1905, when a stable mining proletariat was quickly forming, particularly in Cerro de Pasco, there were 9,651 mine workers in the country.[53] If we add to these numbers the miners' families (estimating five members per family), the population dependent on mining would comprise approximately 4 percent of the whole population of the country.[54] This was a small component of the total national population.

But I have argued that the mine workers in nineteenth-century Peru were also part of the Andean peasantry that went to the mining centers for temporary employment. They were, then, a structural part of the large national and Indian peasantry, although occasionally they would work in the mines. If there was a stable mining proletariat in nineteenth-century Peru, it was only a small part of the mining labor force.

Mineowners, on the contrary, were a substantial part of the national elite—united, at least socially but not always politically, with landowners, merchants, and financiers. Mining created two social classes, mineowners and mine workers, who made their livelihood in mining: one class of entrepreneurs, another of manual workers. (Although, as we have seen in this section, the mining centers and mining in general also implied the existence and agency of larger social groups and other economic and social activities that went beyond these two basic and polar groups.)

But who exactly were these mineowners and workers? To answer this question I first examine the identity of mineowners, particularly in Cerro de Pasco.

Who Were the Mineowners?

In 1827 in Cerro de Pasco a Lista, or Matricula, de los Operarios de Minas y Haciendas was created, counting 60 mineowners and 2,428 workers. Mineowners in Cerro de Pasco were José Apotino Fuster, who had 302 workers under his control, Cesáreo Sánchez with 213 workers, Miguel de Otero with 195, Francisco Goñi with 131, José Casimiro Arrieta with 88, José Lago y Lemus with 78, José Nicolás Lecuona with 77, José de la Cotera with 73, Ramón García Puga with 72, Pablo Minaya with 68, José María de Rocha with 67, and Miguel Maturana with 62. All were owners of mines and *"haciendas de moler mineral"* (mineral-refining haciendas).[55]

In 1878 the principal mineowners, those with the most mines according to the *Padrón general de minas,* were as listed in table 3.6.

The concentration of mine ownership was striking. The six top owners owned more mines (186) than all the town's mineowners of one or two mines (161). The large majority of mineowners in Cerro de Pasco, then, were small entrepreneurs, almost adventurers. Table 3.7 shows this phenomenon more clearly. Bear in mind that if there had been an egalitarian distribution of mine property in Cerro de Pasco in 1878, each mineowner would have had no more than four mines (the actual average per owner was 3.5).

As it stood the fifteen largest owners controlled 47.8 percent of all the district's mines or 320 mines. Six individuals (3.2 percent of the mineowners) owned 27.8 percent of all mines (186), while 55.8 percent of the mineowners, the largest group among them, had 15.7 percent of the mines. Even more telling, 8 percent of the largest mineowners had 47.8 percent of all mines; while at the other extreme 70 percent of them

TABLE 3.6
Holdings of Mineowners in Cerro de Pasco, 1878

Name	Number of Mines
José Malpartida	44
Lagravère and Sons	40
Genaro Maghela	28
Manuel La Torre	27
Jorge Steel and Company	25
Agustín Tello	22
Manuel Chávez	19
Ernesto Puccio	18
Pedro P. Santa María	18
Eduardo O. Villate	16
Mercedes Boni	14
Juan Languasco and Company	13
Félix R. Otero	13
Escolástico Falcón	12
José Aveleyra	11
Total mines of large owners	320
Mines of owners with 3 to 9 mines (40 owners)	189
Mines of owners with 1 or 2 mines (133)	161
Total	670

Source: Padrón general de minas correspondiente al segundo semestre del año de 1878 (Lima: Imprenta del Estado, 1878), 6–11. I have classified and grouped the mines and mineowners that corresponded to the "Distrito Mineral de Cerro de Pasco, Diputación del Cerro de Pasco."

had 24 percent of the mines. A small group of large mineowners and a large group of small ones is the image that could be extracted from the distribution of mine property in Cerro de Pasco in 1878.

If we compare the names of the large mineowners in 1878 with those of 1827, we will see that many had changed. If the Fuster, Otero, and Arrieta families still owned some mines, most of the great mineowners of 1827 had disappeared fifty years later, to be replaced by

TABLE 3.7

Mine Ownership by Category in Cerro de Pasco, 1878

	Category of Owners	Number of Owners	Percentage	Number of Mines	Percentage
Small	I (1 mine)	105	55.8	105	15.7
	II (2 mines)	28	14.9	56	8.3
Medium	I (3 to 5)	31	16.5	124	18.5
	II (6 to 9)	9	4.8	65	9.7
Large	I (11 to 20)	9	4.8	134	20.0
	II (more than 20)	6	3.2	186	27.8
Totals		188	100.0	670	100.0

Sources: Padrón general de minas correspondiente al segundo semestre del año de 1878 (Lima: Imprenta del Estado, 1878), 6–11.

relative newcomers (Malpartida, Lagravère, Maghela, etc.). Historically, the great mineowners, at least in this part of the country, have not retained control of a large number of mines for long periods. They move on to other economic activities, their mines fail, new business groups invest in mine property, and so forth. Mines and mining businesses have changed hands constantly. The instability of mine ownership reflected the nature of mining in nineteenth-century Peru: mining wealth, mining success, depended more on finding new veins of mineral, new ore deposits, in the best case with a large silver component, than on the patient labor of working old ore beds with an ever-increasing capital investment. Thus mining was still a small-scale economic activity with low investments. Mining wealth was based on the ownership of a large number of small mines rather than on a great productive complex backed by extensive capital.

However, the 1870s was a decade of change for mining production and mine ownership. New business people entered into mining, claimed new ore deposits, or bought old mines. These new aquisitions, this new atmosphere, although rather speculative, is reflected in the same Padrón general de minas, where new entrepreneurs such as Ernesto Puccio, Manuel de la Torre, Mercedes Boni, and Genaro Maghela appear as claimers of new mining deposits.[56]

This trend toward new capital is also evident in the nationality of the new mineowners. The 1860s and 1870s were a moment of heavy European and Asian immigration to Peru.[57] Among European immigrants with capital to invest, a favorite investment was mining. Of the six largest mineowners in Cerro de Pasco in 1878, one was French (Lagravère), one Italian (Maghela), and one English (Steel). The remaining three names were Spanish, suggesting that they belonged to the criollo national elite. Other foreign names also appear on the list of mineowners, names such as Myers, Puccio, and Boni. There were more foreign names, however, among the largest mineowners in Cerro de Pasco than among the owners of small and medium-size mines.

Lagravère, one of the largest mineowners, had several years earlier married Emilia Santiváñez, a member of the local elite, with whom he had several children who were co-owners of his mines. Emilia Santiváñez de Lagravère also appears in the *Padrón* as a mineowner. His case shows that marriage, for an European immigrant, was a means of becoming part of the mining upper class. José Malpartida appears to have ended up owning the mines of the Fuster, because in 1878 he was still associated with C. Fuster as the co-owner of two mines. Malpartida is an example of continuity of the 1820s mining elite.[58]

The *Padrón* also shows that, at least in 1878, there were different kinds of institutional mine ownership in Cerro de Pasco. The old pattern of ownership, by congregations or institutions, such as the San Teodoro de Huamico Seminary, was still in place. But there were also some modern incorporated enterprises *("sociedades anónimas," "sociedades por acciones," or "empresas comerciales"),* such as the Compañía La Esperanza, Juan Languasco and Company, or Jorge Steel and Company. In 1878, most mines were still owned, however, by families and individuals, not modern enterprises with shareholders.[59]

The case of Jorge E. Steel represents a break with this tradition. An English immigrant, he brought into the mining sector a new mentality,

more rationalistic and business oriented, not dependent on family relationships or considerations.[60] His properties in Cerro de Pasco were
organized as a real company, a trade house, whose representatives
and managers were Ignacio Rey and M. Gutiérrez.[61] He was establishing a pattern of business organization that two other foreigners,
the Americans John Howard Johnston and Jacob Backus, would later
continue. These latter two founded a very modern enterprise in Casapalca, south of Cerro de Pasco, that would renew Peruvian mining at
the end of the century: Backus and Johnston. The firm of Backus and
Johnston was clearly a shareholding company.[62]

One great difference between family or individual ownership of
mines and ownership by a commercial enterprise is that for the former there was no bankruptcy, either legally or in accounting terms. A
commercial mining company, such as Backus and Johnston, kept a
rigorous account of costs and revenues, in order to calculate its profits.
A family enterprise, on the other hand, was a way to generate income
for the household economy, and thus usually kept loose estimations
at best. When a mine ceased to be productive it simply was abandoned. The family or individual ownership of mines in nineteenth-
century Peru did not function as a modern business, but with rules
akin to those of a peasant economy.[63] Or, in the words of Alan Knight,
it depended on "contacts and clientelism," which "inhibited the 'ra-
tional' pursuit of profit."[64] The great difference, however, between
the mining economy and the peasant economy was that the families
or individuals that ran mining properties in Cerro de Pasco hired Indian peasant labor. There was no self-exploitation, as in a typical Andean peasant household or the peasant economy in general, à la
Chayanov. This is another proof of the artisanal character of nineteenth-century Peruvian mining.[65] This small scale is clearly reflected
in the fact that in 1878 in Cerro de Pasco 105 owners owned just one
mine; and they comprised 56 percent of all mineowners that year.
Among these individuals were several women, such as Josefa Navarro
and Mercedes Boni, owners and perhaps managers of mines. Finally,

many mineowners in Cerro de Pasco and other mining centers were usually absentee property owners. They did not live in the area, nor run their businesses, but gave over their administration to managers or mayordomos.[66]

Mine Workers, Enganche, Wages, Debt, and the Intensity of Labor

Let us turn now to the mine workers, of whom there were 5,071 in 1878 (of which 1,328 worked in Cerro de Pasco),[67] and 9,651 in 1905. I have found extremely rich documentation on their daily work and lives in nineteenth-century Peru. Beginning in 1877, one year after the Lima School of Engineering was founded, students of mine engineering went to mining centers throughout the country to research their economic and technical conditions, as well as their social problems. Over the years a wealth of information was accumulated in their theses and research reports. I have found 112 of these theses and reports up to 1900. The testimonies I have chosen from these documents, although told by middle- and upper-class urban students, are true ethnographic accounts of the living conditions of the Peruvian mine workers at the end of the last century.[68]

About the mines of Huarochirí in 1889, Celso Herrera and Felipe A. Coz report:

> The worker in the mines is the Indian, whose capacity to work is generally known. He is peaceable and obedient, although not all have this character; there are some that can be made to obey through only the most strict discipline. In general he is a hard worker, whose capacity is suitable for a job like that in the mines, and what is useful about this kind of Indian is his constant, always constant, labor. His capacity [for work] and endurance are confirmed by the nine shifts that he regularly puts in each week. He is impassive, he almost never works in order to prosper, and he is not worried about his future.[69]

Notwithstanding the obvious racial and social prejudices in this statement, it is true that "the worker in the mines is the Indian" *(el operario de minas es el indio)*. Although, I should add, it is the Indian peasant, and more specifically the Indian Andean peasant. In a society as socially and racially fragmented as that of nineteenth-century Peru, which was an overwhelmingly rural and peasant society, where the population in the mining centers (mineowners, mine workers, and the rest of the population of mining towns) was, as we have already seen, a minority of the national population, the fact that the mine worker was an Indian means that he or she was not only a member of the Peruvian racial majority, of the large Peruvian Andean peasantry, but also the cultural heir of the old societies and civilizations that existed before the Spaniards arrived in these areas. There is not only an abyss between social classes (mineowners and their workers), but this social and economic division of people in the mining centers was also sustained by cultural and racial divisions.[70]

The other element that clearly appears in Herrera and Coz's report is the extreme hard work required of the Indian laborers. In order to put in nine shifts each week, they had to work daily, from Sunday to Sunday, and still had to work two additional working days in the middle of the week. Twice during the week, then, they had to work consecutively a day shift, a night shift, and yet another day shift without much sleep, probably only chewing coca leaves to fight fatigue.[71]

In Yauli in 1885, four years before the Herrera and Coz research trip, another student, Ismael Bueno, wrote that mine workers there labored from 7 A.M. to 6 P.M. with only half an hour of rest at 9 A.M. and again at 12 P.M.[72] In all, they worked ten hours per day, one more than in Huarochirí, and ten shifts per week. The Yauli workers thus had to put in three *huaraches* (a night shift between two day shifts) per week, in other words, three night shifts. This means that these mine workers had to continue laboring during three simultaneous working shifts of some thirty-three hours three times during the week (working probably Monday, Wednesday, and Friday nights), resting only for a little more than twelve hours three times during the same week, probably on Tuesday, Thursday, and Saturday nights.[73]

The Herrera and Coz report on the mines of Huarochirí also provides the exact point of origin of these Indian mine workers, and the method of recruitment to the mining centers:

> In these days almost all mines, and among them El Rayo, have agents in Jauja, Huancayo, Tarma, with the purpose of hiring people *[enganchen gente]* to work. These agents, or *enganchadores,* working for El Rayo are paid according to the number of people they send to the mines, and according to the number of months the worker has to work. They are responsible if the worker runs away from the mine, although this seldom happens, because it is well known that the mining laws *[las ordenanzas]* state that a worker cannot work in a mine without a card of good behavior from another mine, and if they run away they could lose all their land and possessions [including tools and animals] in the places where they come from. But the El Rayo mine does not depend completely on its agents because every day there are fewer and fewer of them, El Rayo having only two at this point, one in Jauja and another in Huancayo. More than half of El Rayo's workers today are not hired by agents *[no son enganchados]* but work voluntarily in the mine.[74]

In Huarochirí in 1889 and in other mining centers at the time, a migrant population of indigenous peasants came to the mines to work as laborers. However, their migration was not spontaneous; they were not free workers. They had been recruited, "ensnared" *(enganchados)*, by agents of the mining companies (in some cases by independent agents as well) who established themselves where the Indian peasants lived in order to lure them to the mines. It is obvious, as confirmed in the Herrera and Coz report on the Huarochirí mines, that after an initial period of ensnaring these people from rural areas, a current of free workers moved to the mining centers of their own accord. In this sense, the historical period after the War of the Pacific, from 1884 on, would appear to be a moment of transition, from a forced-labor system to a more open, free-labor market, where cash incentives were the main attraction for working in the mines.[75]

Until 1889, according to the Herrera and Coz report, the enganchado peasant could not always run away from the mining centers. A scheme of vigilance was set up in the centers to keep the work going and the forced-recruitment system *(el sistema de enganche)* in place. These restrictions acted as an economic guarantee for a job that was promised to be done. Furthermore, the wage of an enganchado mine worker was reduced in several ways. For example, a deduction was taken to pay the labor recruiter, the enganchador, and one to pay for the worker's subsistence in the mining center. Thus, the mine worker started his new job in debt.

Since the cash advances that the mine agent gave to the potential worker were guaranteed with the peasant's possessions, in accepting the job, and especially the cash advances, the worker was also putting his belongings at risk. Is it possible to conceive today that the labor performance of a worker could endanger one's belongings, one's land, one's home? The enganchado peasant stood to lose his land in his place of origin, where the enganche contract was signed.[76] We see here, then, a mine worker who did not go freely to the mining centers to look for work. He was tempted and later recruited by a special agent of the mining companies. He was sucked in, attracted, motivated, forced, to go to the mines.[77]

In 1891 another mining engineering student, Francisco R. del Castillo, made a similar comment on the workers of Yauli and Huarochirí:

> They are almost all from the Jauja province; only a few are from the same area [as the mines]. The latter prefer to work on their plots of land *[el trabajo de las chacras]* and those who work in mining usually are not good workers; they are almost always lazy and do not do the job well because of the limited practice they have. . . . In the mines of Casapalca too almost all the workers are from Jauja; they work for a daily wage or to complete a specific task *[por jornal y por contrata]*,

putting in a huarache only once a week. The Esperanza workers are paid in cash wages every fifteen days and it is impossible not to give them cash advances. In the mines of Casapalca the workers are also paid every fifteen days, not only in cash, but also in goods and in foodstuffs, these last forms of payment being called *acomodana*.[78]

The observations of del Castillo, and especially the one I quote here, give us much information on the forms of payment of the migrant mine workers *("los trabajadores mineros enganchados")* in late-nineteenth-century Peru. There was not only the problem of money advances and the question of enganche, the coactive—or if we prefer the softer term—, the induced recruitment of workers, but there was also the question of the workers laboring on a day shift and paid in cash *(por jornal)*, or hired for a specific task and paid after its accomplishment *(por contrata)*.

The workers of Casapalca furthermore (and this is clearly pointed out in del Castillo's report) are paid in cash but also in goods and in foodstuffs. Not all the miner's salary is a free-money wage. Wages were often dependent on other social and economic considerations, such as the kind of labor arrangements (por jornal or por contrata) and the forms of payment. In the case of the payment in goods or foodstuffs, the employers certainly had a clear advantage in that they could buy the goods, or produce them, at low prices and sell them to workers at high prices. In this case the employer, the mineowner or *el mayordomo de las minas,* obtained an extra surplus, lowering his or her labor costs even further, while the mine worker saw his or her real salary reduced.

It is not surprising that mine workers were paid every fifteen days in a cash wage arrangement, as was the case partially in Casapalca or in the La Esperanza mine. This modern form of labor contract still rules. But if they worked por contrata it meant that they were paid in ores they extracted from the mines, while buying their own tools and equipment (candles and metal bars to dig in the tunnels). This is exactly what Ismael C. Bueno explained in 1887, when he was an engineer and director of the Escuela de Capataces de Cerro de Pasco (Cerro de Pasco school for foremen):

> Although mining wages were fixed on 11 January 1887 by the Diputación de Minería in agreement with the mineowners, those wages are valid only in part, and we could say that in a day the barretero [mine digger using iron bars] earns 0.40 to 0.50 soles per day. More common however is taskwork *[el trabajo por tarea]*, in which the laborers have to extract a *cajón de medida* [20 quintals] of ore, which is a third of a *cajón de ley* (60 quintals). The wage in this case goes from 0.60 soles to 2.50 soles per *cajón de medida*, according to the ease and the underground path taken to extract the ore. In this case all the expenses are carried by the barreteros, but they have the right to extract a certain amount of mineral, which is called the *ración*.[79]

On one side there is the law, the rulings of the Diputación de Minería; on other side is daily economic reality. The Diputación fixed some wage levels; in reality workers were paid at a lower level. Furthermore, these wages were paid not por jornal, as a daily cash wage, but por tarea, per task. The mine worker, the barretero, paid the expenses associated with the work, but was allowed to extract some silver for himself.

Recall that Herrera and Coz reported that in Huarochirí in 1889 mine workers had to work up to nine shifts per week. From their report and Bueno's discussion of the barreteros, a picture emerges of the extreme intensity of labor in the silver mines in late-nineteenth-century Peru, as well as an absolute exploitation of this manual worker, of this Indian (to emphasize again the pejorative racial label applied to the Peruvian peasant mine worker).[80]

Two additional reports, from 1892 and 1894, complete our picture of the social working conditions of the late-nineteenth-century Peruvian mine worker. The first, by Julio A. Morales, also describes these indefatigable mine workers recruited by enganchadores far beyond the mining centers:

The way these mines obtain their necessary workers is through the enganche system. This is but the commitment that the workers agree upon in order to work in the mines for a period of time, for a sum of money that is advanced to them in the place in which they are recruited. The majority of these workers are from Tarma and Jauja, where there are people in charge of carrying out these enganches and who are paid on commission. The workers are strong and accustomed to mine labor. Although they do not have a great commitment to this kind of work, we could not say that they are bad. Their defect is that they like *charta,* rum mixed with water, coca leaves, and quinine in an infusion. As regards their moral state, they are good people, although in general they have no education.

Further on Morales notes:

The way to pay them is part in money and part in goods. At the Eliza mine they are given acomodana every fifteen days and they are paid [in cash] every month. In Aguas Calientes they are given acomodana every Sunday and they are paid every month too. The boys who work screening ore or as *carreros* [pulling mining wagons] earn from 25 to 30 centavos per day. The acomodana is but the goods they take from the store; they are also given cloth. In this way each mine has a store with everything that is necessary in those places and from which the workers are provided with everything.[81]

The reference makes it clear that a large part of the miners' wages was paid in goods through the company store, where the workers bought hard liquor, coca leaves, quinine, foodstuffs, even clothing. In 1894 Santiago Marrau reported:

These days the mine has 150 workers, three *caporales* [foremen], one *mayordomo general de minas* [mine overseer], and one *mayordomo de cancha* [yard overseer]. The workers, who usually come from the Jauja valley (Muquiyauyo), are hooked [enganchados] by special people called enganchadores, through an advance of money whose goal is to motivate them to go to the mines. Among these workers some, as I said before, are in charge of the groundwork; these are the barreteros. Others carry the mineral ore from the production front to the wagons that take it to the mine surface; these are the apires; and finally the carreros, mostly old people that move the wagons. . . . The workers labor daily from 7 A.M. to 10 A.M. and from 10:30 A.M. to 6 P.M., taking another short rest at 2 P.M. Furthermore, three times a week they have what they call *guaracha ("huarache"),* that is, the night shift. It was only later that the mine adopted the system that uses two sets of workers, one for the day shift and one for the night shift.[82]

Thus, mine workers had very intensive work days that were ten and a half hours long, in addition to working three nights per week. Their recruitment was in some cases forced or induced. Their wages were paid in kind or in goods bought at the company's store, putting them sometimes in debt. A grim picture results from this study of the lives and labor of Peruvian mine workers at the end of the nineteenth century. However, as it has been shown in a previous section, these peasant workers were just temporary miners. They went to the mines to obtain the necessary cash for their expenses in an increasingly commercialized economy. The hard working conditions were bearable enough because mine work was just a temporary occupation, and was paid in money. Let us now return to the economic and physical context of mining.

In the nineteenth century the mines of Peru were in the Andes, thus the export of minerals and the import of supplies for mining operations depended on transportation between the mountains and the ports on the coast. A trade network had to develop to articulate the mining industry to the rest of the national economy. Trade and transportation, therefore, will be the focus of the next two chapters.

Mining Center of Tarica, in the highlands of Ancash.

Mining Center of Casapalca, in the highlands of Lima.

A muleteering stop in the Andean highlands.

Railroad from Arequipa to Puno.

Mining Center of Caylloma, Arequipa.

Customs of Arica.

The port of El Callao.

Chapter 4

Transportation and Trade

Merchants, Muleteers, and a Railroad to Come

> ... Y prendido a la magia de los caminos,
> el arriero va.
> (... And captured by the magic of roads,
> the muleteer goes by.)
> —Atahualpa Yupanqui, Andean folk song

MINING WAS GEOGRAPHICALLY dispersed in nineteenth-century Peru. If there was one region of relative concentration, it was Cerro de Pasco, which dominated Peruvian mining throughout the century, although the sierras of Cajamarca, La Libertad, Ancash, Lima, as well as Ica, Arequipa, and Puno, were also mining regions. But mining regions were also economic regions in a broader sense. Here the mining industry was articulated with other economic activities—for example, agriculture, livestock herding, and crafts—through trade and other forms of commodity circulation and exchange. This articulation reflected the fact that the product of mining was a commodity, preeminently silver, that had to be exchanged for other goods and services. Silver was in part an export commodity, destined for foreign markets; but it was also a commodity for domestic use and a medium of exchange for the internal market. Thus the silver obtained from Peruvian mines was exchanged through internal commerce with other goods produced in the regional economy. And when any mining products (silver, gold, copper, etc.) were exported, they required an internal transportation system to reach the various ports of the Peruvian coast. In the case of the domestic exchanges within the internal market, the mining commodities circulated through different towns in the interior of the country using the means of transport prevalent at the time, muleteering.

The transportation of minerals and other commodities in the interior of the country was done by mule or llama teams; the people involved in this trade were called respectively *arrieros* and *llameros*. But they were not just carriers of the mining or other goods involved in this fluid network of exchanges. They were also merchants, traders, small or large entrepreneurs, including sometimes even small campesinos, who profited from this business, from this coming and going of ores, metals, and other commercialized goods.[1] Arrieros and llameros were partially displaced after 1850s, and more so after the 1870s, when a railroad network began to be created, changing the nature and circuits of these exchanges, of these markets. With the rush to build railroads during the administration of President José Balta (1868–72), the locomotive and the railroad track started to be a common feature in the transportation networks of nineteenth-century Peruvian mining. Before then, for the first seven decades of the century, muleteering *(arriería)* connected mining with other economic activities of the country, transforming the mining regions into regions of full economic participation.[2]

The muleteering system acted to more strongly articulate domestic markets and internal circuits of transportation and trade, creating backward linkages with the whole national economy. The railroad, in contrast, opened the mining economy more freely and efficiently to export markets, and therefore to a more direct influence of the forces of the international market and capital. Thus, the 1870s started

THE BEWITCHMENT OF SILVER

a period of transition, when muleteering and the railroad coexisted but were in conflict with one another. Focusing on transportation and trade, I will open a new dimension to look at the social economy of nineteenth-century Peru.

Problems of Transportation and Trade in Early-Nineteenth-Century Peru: From Mining Regions to Regional Markets

To recapitulate, I have shown that silver mining, and to a lesser extent gold, copper, tin, mercury, and coal mining, together with the mining of other minerals, constituted a substantial economic activity in nineteenth-century Peru. About 2,000 mines yielded these minerals, and these mines were owned by 700 mineowners and worked by 9,000 mine workers. Finally, the mining took place in various geographic regions of the Andes, especially in Cerro de Pasco, but also in the sierras of Lima, Ancash, Cajamarca, and so forth. Each of these three elements had, of course, a specific evolution, with its own fluctuations and changes throughout the century, but the tendency is rather toward continuity than change.

Hence large amounts of minerals had to be transported from the mining centers to internal markets and to the ports. In 1897, for example, the British consul in Peru, Alfred St. John, described in a report to the Marquis of Salisbury the transportation of minerals from Cerro de Pasco, the sierras of Lima, Ancash, and Cajamarca, to several Peruvian ports for export. He made clear that this output of Andean mines in several areas was finally commercialized internationally and traded in several world market centers.[3] In the 1896–97 fiscal year the value of these "Peruvian mining exports," according to the report, reached approximately 11,000,000 soles. St. John noted that these exports were carried out by the Central and Southern Railroads, and that the "railroads, as well as the steamships, are kept in a very good state."[4] However, he also mentioned roads and highways that were being built or maintained for the transport of mining commodities as well as other "goods of trade." The consul dreamed of a day when the Peruvian road system would connect the Pacific coast with the Amazon River and thus eventually the Atlantic shore.[5]

Large amounts of ores and metals were being traded and transported throughout the country, particularly to ports for shipment overseas (see my data in chapter 2 on silver, gold, copper, and tin exports). Thus, clearly the mining economy of nineteenth-century Peru was an export economy, but that export sector depended first on an internal network of transportation and trade. Therefore, exempting perhaps the scanty production of mercury, coal, lead, and iron, a great result of Peruvian mining production in the nineteenth century was international commercialization, through which an internal economy was closely linked with an external one (that of Europe and the United States, of the merchant houses of London, Hamburg, Paris, New York).

The theory of dependency continues to have a strong influence in the Latin American social sciences.[6] My goal in this chapter is to reconstruct the internal functioning of the mining economy in nineteenth-century Peru, which only later became articulated with the international market and with dependency mechanisms. By *internal functioning* I mean the internal exchanges, the domestic market, within which mining operated.

The domestic dimension, the internal face of an export sector, emerges most clearly in the case of those nineteenth-century mines located deep in the Andean sierras. Their articulation with the world market, their external articulation, would depend on two elements that expressed a complete network of economic and social relations, two social groups different from the mineowners and mine workers we have studied in the previous chapter. The first, transporters and carriers of mining goods, brought minerals out of the mining centers and also carried supplies and equipment used in the mines. The second, merchants and traders, were also essential actors in the making of the mining economy in nineteenth-century Peru. These two groups traveled over a domestic road system and into and out of ports that were necessary for the development of their businesses. Thus, roads and ports were expressions too of the internal transportation system and the external trade of the mining economy in nineteenth-century Peru.

I have estimated the value of silver production for the whole of the nineteenth century at more than 300 million soles, and we have already seen that copper started to become competitive with silver in the second half of the century (particularly between 1861 and 1864, but mostly at the end of the century).[7] Add to silver the value of gold, copper, and other minerals, and we have an enormous mineral wealth that had to be moved from mine to port through the internal road system. As much as 350,000 marcs of refined silver per year (some 100,000 kilograms) were moved constantly throughout the country. The main connecting port for the Cerro de Pasco mining economy was Callao (the port of Lima), while for Hualgayoc the silver trade and the export business passed through the city of Trujillo.[8] For the Cerro de Pasco–Lima–Callao connection, the road through Obrajillo and Canta was the main transportation route, but there was also an alternative exit route through Cajatambo to the port of Huacho.[9]

Yet things were not so simple. The transporting and trading of volumes of ore and silver bullion from mines to ports across the routes and roads of the mining commercialization system entailed the interplay of various interests, but particularly the prevailing interests of silver merchants and mineowners. The merchants controlled the exchange, commercialization, and financing of the mining business, whereas mineowners only controlled production. One document from the early national period reveals in some detail the different aspects behind this relationship between merchants and mineowners, between traders, transporters, and producers of Peruvian minerals. I will focus on Landaburu's report on these matters because it is an excellent entry to understanding the characteristics of domestic mining trading in the first half of the nineteenth century.[10]

In 1827, in a report addressed to the National Congress, Juan José Landaburu commented on problems in the transportation and marketing of minerals. His report was part of a broader debate about the wisdom of creating official *bancos de rescate* (marketing and credit banks).[11] The bancos de rescate would have reduced the control of private merchants over mineowners in the marketing of mining goods and supplies at the beginning of the national period.[12] Silver merchants based in mining centers, such as Cerro de Pasco or Hualgayoc, or in commercial towns linked to these mining centers, such as Trujillo or Lima, bought silver metal at low prices, and issued high-interest credit to the mineowners—the mining entrepreneurs directly involved in the extraction of the minerals. These merchants also supplied the miners with mining inputs, consumer goods, and other *habilitaciones* at exorbitant prices. Thus Landaburu's description shows the difficulties mineowners faced in gaining access to the transportation and exchange networks, and their consequent dependence on merchants and financiers.[13]

Landaburu wrote from Lima, but he demanded trade concessions that would make it easier for minerals extracted in La Libertad mines to reach the port and market of Lima. One such concession was that mine products could be transported by sea. The following passage from his report compares the commercial situations in Pasco and La Libertad, two of the most important mining regions in the nineteenth century:

> It is true that the Banks of the Department of La Libertad should pay a different price for the [silver] bullion than that paid, or promised to be paid, in Pasco. It is good to consider that in this latter mining center, more silver is extracted, it is easier to buy, collect, and send the bullion to this Capital [Lima], just as it is easier to return the capital invested in these purchases. When all these operations are harder

and three times longer, as in the places where Banks are proposed to be established, in that Department [of La Libertad], then the [mining] entrepreneurs are deprived of the interest and benefits that these invested funds should produce. It is for this reason that the permit to send the minerals by sea is demanded, because the distance of some two hundred leagues [about 1,000 km.] not only makes transportation take longer at exorbitant costs, but there are also imminent risks of Robbers, Rivers, and others. The necessary financial guarantees could be made in Trujillo, from whose port the bullion is put on board, or they could be presented at the Lima Mint, in this Capital. These financial guarantees could also be presented in this principal Custom where the [mining] goods produced in the Department of La Libertad enter.[14]

The passage shows several problems in the trading and transporting of mining goods, particularly silver bullion. Where there was greater extraction of silver, as in Cerro de Pasco, it was easy to "buy, collect, and send" the minerals to the trading town, in this case Lima. This also facilitated the return of capital to be invested in mining production and this greater access to funds allowed mining capital to complete its productive cycle. Thus, some complementarity existed between producers, merchants, and transporters. There was less conflict between the interests of all these three actors, and a larger fluidity existed in the exchanges of money, capital, and mining goods.

But where there was less silver extracted, longer distances, and higher transportation costs, as in the department of La Libertad, between the mining center of Hualgayoc and the city of Trujillo, the return of capital was limited, its access more restricted, and the mining productive cycle was not completed. The mineowners, the people in charge of production, thus did not realize all the benefits from the sale of the mining goods. The traders and transporters played a larger role and kept a larger share of the profits. Thus, to encourage mining production and growth—and this is the most important recommendation of Landaburu's report—a good fit was necessary between trade and production, financing and production, and, finally, transportation and production. Marketing, finance, and transportation were the bottlenecks in the development of mining in nineteenth-century Peru, at least in this early period.

The same observation is made in a report of the Tribunal de Minería, the official administrative body and court of the mining sector, a few years later, in 1834. The members of the tribunal wrote that "the same financiers and traders *[aviadores]* that are in charge and take the risks of that business *[esa especie de giro]*, far from helping the mineowners, in reality cause their ruin."[15] The marketing and transportation of mining goods, the dominance of merchants and financiers who supplied the mining centers *("habilitaban la minería")*, and their hegemony over the mining producers, continued to be a major obstacle in the development of the mining economy throughout the century, even after a major technological innovation in transport, the building of railroads, had taken place. But let us not rush ahead, but pause and observe the mining commercialization system, this trading and transporting of minerals and other goods, in the early part of the national period.

By the time ore reached the refining mill *(ingenio)*, the first transportation process had already taken place. Carrying the hundreds or even thousands of pounds of silver ore, for example, from the tunnels (bocaminas) and socavones where it was extracted to the center where it was processed into metal was the job of the muleteers or, more precisely, llameros. In 1830, for example, *cascajos* (gravel, mineral nuggets) from the Carmen, Registro, and San Juan de Dios mines were processed and carried to the Tinyahuarco ingenio, one of the largest in Cerro de Pasco. Between June and August of that year, 255 pounds of refined silver were obtained in *circo* (a flat processing area) no. 1 from eight *cuerpos de cascajo* from the Carmen mine, eight from the Registro mine, and eight from the San Juan de Dios mine. In the period leading up to 28 July, 120 pounds of refined silver were obtained in circo no. 2. Finally, around the same time, between June and August 1830, 134 pounds were obtained in circo no. 5, 122 in no. 6, 120 in no. 7, 153 in no. 9, 90 in no. 10, and another 90 in no. 11.[16] All 1,084 pounds of silver extracted from more than 9,000 pounds of ore nuggets were carried by 24 mules and 47 llamas. The llamas, the Tinyahuarco books report, did not walk more than three to four leagues per day (15 to 20 km.). Thus, scores of mules and llamas, carrying thousands of pounds of cascajos, created a direct link between the Carmen, Registro, and San Juan de Dios mines and the Tinyahuarco ingenio.

The carrying of ore to the official smelting house *(la casa de fundición)* was a different matter. In Cerro de Pasco, the official smelting house was located in the village of Pasco, several kilometers south of the *cerro rico* of Colquijirca, the core of the mining center, and past the *estancia* (ranch) of Llacsahuanca.[17] Trains of llamas, and probably also of mules, had to go from the different ingenios to Pasco carrying the pieces of refined silver that would be smelted into silver ingots. Most of the official smelting houses throughout the country did not survive the 1830s. Only the casa de fundición in Cerro de Pasco flourished throughout the century, proving again the predominance of this mining center in the central sierras of Peru.[18]

In two final stages the ores, now transformed into silver ingots, were transported from the casa de fundición, or *callana,* in each mining area of the country at least until the 1840s, to one of a number of mint houses and, then, to the ports for export. Mining goods, particularly silver, were in constant motion, from mine to ingenio, from ingenio to casa de fundición, from casa to mint, from mint to port. There were then four stages in the transportation of silver from the mine to the port:

(1) The silver ores extracted from the mine shafts had to be carried by mule or llama to the ingenio, where they were refined into silver metal and shaped into pastes in the form of pineapples *(plata piña).* In 1858, for example, a calculation was made

of the "costs and benefits" of processing six boxes *(cajones)* of ore in Cerro de Pasco. If the extraction of these cajones cost 78 pesos, their *bajas* (transportation from mine to mill) was estimated at 54 pesos, an amount only 31 percent less than extracting the ore from the bowels of the earth. Transportation also represented 21 percent of the total costs of extracting and processing these six cajones of ore.[19]

(2) From the refining mill, the plata piña was taken to the official smelting house (callana),where it was transformed into silver ingots, each stamped with an official seal. The data provided by the British consul, Alfred St. John, show that in the 1896–97 fiscal year slightly more than one-third (3,980,000 soles, or 38 percent) of all silver exported was in ingot and piña, another third (3,500,000 soles, or 34 percent) was in sulphurated silver, and almost another third (2,950,000 soles, or 28 percent) in silver ores *(minerales argentíferos)*.[20]

(3) From the smelting house, the ingots were transported to the mints in Lima, Cusco, Arequipa, Trujillo, and Pasco, where they were transformed into silver coins.[21]

(4) Finally, the coins were taken from the mint house to the ports to pay for imports arriving there or to different marketplaces where international transactions took place, or to leave the country as outflow capital.

Silver exports sometimes varied in form. Between 1825 and 1840, for example, silver was exported primarily as coins (three to eight million pesos each year).[22] Later in the century, however, exports of money decreased and even stopped. This was true during the guano boom, between 1845 and 1876, when the export of silver as a raw ore or processed metal became more common than the export of silver coins. Of course the silver exported as raw ore or processed metal did not pass through the mint house; sometimes it did not even pass through the smelting house *(callana de fundición)*. During the boom, guano exports paid for Peruvian imports. Therefore, silver did not need to be coined within the country, and could be exported raw, as ore.

At the end of the century, after the War of the Pacific, several private smelting plants were set up in mining regions, such as Casapalca or Cerro de Pasco. These plants began to process raw ore into metal, and not, as before, to transform refined silver into official silver ingots. Two stages had been eliminated from the transportation cycle; silver ore moved from the mines to the private smelting plants, and then directly to the ports. A report of 1895, for example, described the *oficina* of Humanrauca, near Cerro de Pasco, made up of "one reflector furnace of two sections, two reflector furnaces of one section, one spout furnace, two cupellation furnaces, one horizontal engine, and one large fan" to cool the whole process. This plant sent refined minerals directly to port using the central highland railroad system.[23] The four stages of the early nineteenth century (from mine to ingenio, from ingenio to callana, from callana to mint, from mint to port) were now reduced to two (from mine to smelting plant, from smelting plant to port) at the end of the century. The railroad and technologically advanced smelting plants changed the networks and the nature of the early trading and transporting of mining goods. Mules and llamas were displaced by steam locomotives and railroad tracks. Mining based on the production and marketing of precious metals also became more industrialized. And silver was replaced by copper as the dominant mining product.

The following examples illustrate the changes in the production and marketing of precious metals over the century. In 1844, of the 439,082 pesos of "gold and silver minerals" exported from the port of Islay in Arequipa, 145,567 pesos (33 percent) were in gold and silver coins, the rest (67 percent) was silver in ingots (that had been stamped in the callanas), old silver-plated craft items, and gold dust.[24] In 1878, at the end of the guano boom, exports in gold bullion were valued at 712,660 soles, silver bullion at 4,119,540 soles, silver ore at 841,934 soles, and stamped gold and silver (meaning gold and silver coins) at 346,829 soles.[25] Little coin was exported in this year; it amounted to 5.7 percent of the total mineral exports, as against exports of raw ores (25.8 percent) or processed metals (68.5 percent). This same year (1878) shows that copper ore was transported directly from the mining center of Cauza in Ica to the ports of Pisco and Chincha without much intermediation or processing.[26]

Most silver mined in Cerro de Pasco went to Lima, not more than "52 leagues" (some 250 km.) away.[27] This proximity meant that profits from silver sales in Lima were likely to make their way back to Pasco to be reinvested in the mining sector. Other producing regions were more isolated from their markets. According to Landaburu in 1827 the transportation and commercialization cycle was three times longer and more difficult in La Libertad than in Pasco, and since, in such regions, the owners failed to receive the benefits of invested funds, he deemed it necessary to set up several bancos de rescate in order to facilitate the mining business.[28]

Landaburu, like other observers,[29] pointed out factors that hindered the development of the transportation and marketing systems in mining centers other than Cerro de Pasco: the long distance from Lima, annoying delays, excessive costs, and the mountainous topography to be traversed. The route from mine to market also contained other dangers, particularly robbers. There were also financial obstacles, such as the need for insurance and other financial guarantees that the goods be transported. It was these obstacles that led Landaburu and others to propose the creation of government-sponsored bancos de rescate. Landaburu, as mentioned, also proposed allowing the transportation of silver by sea, an idea reiterated by the intendant of Chancay in 1826, who also proposed allowing the transportation of minerals by sea—although in his case it was coal.[30] The problem with allowing free sea transportation, according to officials of the Peruvian government, was twofold: first, it would facilitate the already existing contraband trade, and second, merchants would export minerals to the international market directly, instead of trading domestically, along the coast. By trading abroad, they would evade further export controls and the coinage of silver and gold metals.[31]

Complaints about the harsh topography for the transport of mining goods and supplies were often voiced in the early national period, when muleteering was the norm and the railroad still just a dream. In 1828, for example, the second magistrate *(el segundo vocal)* of the Junta Subalterna de Minas in Lampa, department of Puno, lamented "the great damages that the miners suffer because they have to go to the city of Puno in search of gunpowder for their mining operations, and in addition to being twenty long leagues away, they have to cross two very dangerous rivers."[32] The document suggests a clash of interests between different mineowners in Lampa and Puno, with the Lampa miners resenting the trip to Puno to buy their supplies.[33] To have to travel a hundred kilometers for the gunpowder that was indispensable for working the mines greatly increased their operational costs. Added to this were rural banditry and the treacherous Peruvian roads, topics described by authors such as Alberto Flores Galindo, Carlos Aguirre, and Charles Walker,[34] as well as in the sources I found. Toward 1840, for example, one observer noted that the military escorts that went with the silver loads from Cerro de Pasco were "not always capable of resisting the assaults of the black hordes."[35]

The Landaburu report, then, is a good testimony of the different nature and characteristics of the various mining regions, and mining markets, in nineteenth-century Peru, where the roads were not safe, the topography complicated, and the conditions for the transportation and commercialization of mining goods, particularly silver, difficult.

Thus, muleteering, the traditional system of arriería, was a crucial business, as well as a social and economic activity that imposed additional costs on mining. In some ways transport and marketing were at least as important as the extraction and processing of the ores. More than just the difficulty of transportation, profit from mining was also a question of location, of the availability of capital and of economic resources in general, of having the means, the mules, and the money to get to the markets. These were the reasons why the roles of the muleteer and merchant were so relevant in the making of the mining economy in nineteenth-century Peru. Some, particularly merchants, were very much interested in preserving their key role and in charging additional costs to the mining economy, so they could maintain their dominance over the mineowners.[36]

On the other hand, among the muleteers, or arrieros, there were also llameros, transporters who were part of Indian peasant communities, who used their own native animals to contribute, and of course profit, from the transporting of mining goods.[37] These llameros were mostly concerned with the bajas, or the carrying of ore down from the mines to the refining mills.[38] But in general the arrieros were part of a rural middle-class group who tied their activities to those of larger merchants and merchant houses, which in turn were tied into commercial networks throughout the country. In the early nineteenth century, as another historian has shown, *"arrieros muleros"* were using two to three thousand mules in Cerro de Pasco for daily transport within and beyond the mining center.[39] Let us see now how the arriería system operated in the early nineteenth century.

In 1836, well before the arrival of the railroad, Domingo Olivera, from the *Alcaidía* of the Lima mint, informed the treasurer of the mint that:

Into this office under my supervision silver bars have been escorted under the charge of the following:

Don Federico Feyfar	2 bars
Don Antonio Negrete	1 bar
Don Juan Ugarte	1 bar
Don Francisco Quiroz	1 bar
Don Manuel Alvarado de la Torre	1 bar
Don Pascual Bieytes	1 bar

There are 7 bars in all.[40]

Olivera later adds that Meregildo Pies had received "fourteen pesos as payment for the transportation of the seven bars mentioned above."[41] It is almost certain that this arriero led two mules in harness to carry the seven bars, with an approximate weight of 1,190 marcs and a value of 9,520 pesos, from Pasco to the Lima mint.

The six merchants listed in the document shipped the bars on consignment *(a consignación)* and would later pay the mineowners for this silver out of the proceeds from its sale at the mint. At the end of the trip the arriero Meregildo Pies received his fourteen pesos from the merchants. A chain was thus established between the merchants, the person in charge of the transportation of the silver bars, and the actual carriers. I have found other instances of the transportation and marketing of silver bars from Cerro de Pasco to Lima in 1826 and 1843, before and after Meregildo Pies's trip, which show the same pattern.[42]

In May of 1836, another group of muleteers transported loads of silver bars from Cerro de Pasco to Lima: Pies, Vicente González, and Fermín de la Cruz. De la Cruz was most explicit about how he was paid, declaring he had received "eighteen pesos for the nine bars registered on this list," and on another occasion, "sixty-six pesos for the thirty-three bars [or] two pesos per bar."[43] The proceeds to the muleteer were registered as "the gratuity for the driver" *("la gratificación al conductor")* and, as before, were the equivalent of two pesos per silver bar transported. In the four deliveries that the documents for 1836 show the muleteers received 34, 18, 66, and 14 pesos, or a total of 132 pesos.

Let us pause to reflect on the relationships established between merchant and muleteer, or between middleman-financier and transporter, as portrayed in the documents discussed above. Meregildo Pies first brought seven silver bars from Cerro de Pasco and was paid 14 pesos, two pesos per bar, the same fee paid later, in May 1836. In a later trip, Pies, Fermín de la Cruz, and Vicente González brought sixty-eight bars and were paid 84, 34, and 14 pesos, respectively. The first seven bars were worth 9,520 pesos; the remaining sixty-eight bars, 92,480 pesos, a huge amount of money. The payment to the muleteers was one-tenth of one percent of the total value of the bars. The cost of transporting the silver bars, then, was not in the payments made to the actual carriers, but in the merchants' profits. Of course, not all the money the merchants received was profit; they had to pay the mineowners for the minerals and certainly had other costs as well. Still, between the value of the goods transported, particularly silver bars destined to the Lima mint, and the payment made to the muleteers a huge gap existed. The muleteers, however, did not fare badly: Fermín de la Cruz made 84 pesos for just one trip from Cerro de Pasco to Lima. Several such trips would have provided him with good earnings, and, therefore, with a comfortable living.

It is likely that the relationships between merchants and muleteers involved a high degree of trust, of clientelism, and even of personal dependence. The muleteer was the actual carrier of thousands of pesos of silver bars belonging to merchants in Lima or Cerro de Pasco (the Feyfar, Negrete, Ugarte, Quiroz, etc.). The merchants sustained these networks, in which the muleteer played only a small part. The scope of these networks, however, depended on the nature of the trade and, more important, the nature of the goods traded. In the case of silver bars, a commodity of high commercial value, trust and personal dependence between merchant and muleteer were even more significant. For an early nineteenth-century Peruvian mining merchant it was key to have as broad a network of trustworthy persons as possible.

That the silver bars indeed came from Cerro de Pasco was confirmed in a communication dated 2 May, as well as in one on 15 May: "I am sending the list of silver bars that have come escorted from Cerro de Pasco."[44] The phrase "have come escorted" is noteworthy: one could not send silver bars with a commercial value of thousands of pesos without any protection. Guards therefore escorted the deliveries. These guards often were hired in Lima and sent to await the mule train in Llangas or Santa Rosa de Quives, forty or fifty kilometers from the capital. Before that, the mule train would not be attacked by road robbers because the route wound through a narrow canyon in the sierras from Canta or Obrajillo to Llangas. At both ends of that canyon *(quebrada)* were towns where the local authorities could muster armed deputies. But between Llangas and Lima, on the open coastal plains and valleys attacks were frequent. Here the bandits, most of them Afro-Peruvian runaway slaves *(libertos)* could find easy refuge.[45] Rural banditry was so widespread at the end of colonial times and in the early republic that, as historian Alberto Flores Galindo notes, the capital of the country felt at times like a city under siege.[46]

Once the mule train entered Lima it first went through city customs ("silver bars that have entered this Customhouse coming from Cerro de Pasco"), and then to its final destination, the Lima mint. Every time the merchants, now the owners of the silver bars, are mentioned in the documents, they are said to have the silver bars "on consignment" *(en consignación)*, meaning in this case that the mint would not pay for that delivery of silver until it had been physically transformed into coins. In the meantime, the merchants in Lima or in Cerro de Pasco were not paid, and were in effect giving credit to the Lima mint for a period of time: one week, several weeks, a month. We have to suppose that these credit relations, this delay in the payment of the silver delivered, lengthened the period between when the mineowner turned over the silver on consignment to the merchant and when the merchant paid him or her. Mineowners thus in effect extended credit to merchants during these delays. Hence the mining economy involved significant credit and relations of trust. In sum, the movement of capital from the mine to the mint and back again was slow in the first half of the nineteenth century.

I have found documents similar to those described above, also from the 1830s but concerning other regions of the country. They attest to the existence of different regional mining markets and different market logics, as Landaburu had contended ten years earlier. Between 5 April and 27 July 1830, sixty-seven silver deliveries belonging to eight merchants, entered the official smelting house of Trujillo to be transformed into silver bars.[47] The original source of these bars, according to other documents, was the mines of Cascabamba, Pienvre, Quipa, Tembul, Collocedas, Chilete, Yucual, Retama, Sotonpampa, Succhabamba, Trinidad, Totuguayco, and Lambar—all in the provinces of La Libertad.[48] There were eight merchants trading silver bars between various mines in the provinces of La Libertad and the city of Trujillo in 1830; and twenty-four merchants handled about the same amount of silver between Cerro de Pasco and Lima in 1836. Some Cerro de Pasco merchants had only a single bar on consignment, whereas among the Trujillo merchants the smallest delivery was of three bars. And one Trujillo merchant (with a foreign name, Henrique Barnad) had seventeen bars worth about 23,120 pesos, a large sum of money.[49]

It is true that the Trujillo documents span four months, so the sixty-seven silver bars entered Trujillo at a rate of almost seventeen bars per month, whereas in Cerro de Pasco the documentation covers only one month. The sixty-eight bars that arrived in Lima in May 1836, then, is the monthly figure, four times higher than Trujillo's. This shows once again the larger dimension of mining production and trading in the Cerro de Pasco area. In the provinces of La Libertad and the city of Trujillo, a smaller mining economy produced a larger concentration of merchants in the distribution component of the mining economic cycle, whereas in the Cerro de Pasco–Lima circuit a larger mining economy resulted in a lower concentration at the transportation, commercialization, and distribution levels.

The foregoing has provided a picture of merchants and muleteers in the making of regional markets spawned by mining. I now present a general picture of the Peruvian internal trade in the first decades of the nineteenth century, and study with more detail one of these trader-muleteers, Patrick Guinnes, or Patricio Ginez as he was called in the documents. Before the coming of the railroads in the 1850s, internal commerce depended almost exclusively on muleteers, on arrieros. Arrieros transported ore from the provinces of La Libertad to Trujillo, from Cerro de Pasco to Lima, from Lampa to Puno, from Caylloma to Arequipa. Arrieros, whether mestizos or Indians, were constant travelers on the web of roads that made up the internal market and communication system that connected mines with market towns and ports.

Internal Trade and Regional Articulation: Foreign and National Merchants, A Credit Economy, One Arriero, and Many Mules

In 1828 the French consul in Lima, informing his government about a new bill that the Peruvian Congress planned to approve, wrote that the foreign merchants were in stiff competition with the natives *(los naturales)* in the trade of goods from the province of Carabaya in Puno. Carabaya was a Peruvian province that produced minerals such as gold and silver (more of the former), as well as quinine. The Puno representatives in the national congress had criticized the foreign merchants "who come to the mountains of this province," and wanted to approve some regulations to force these merchants "to buy the goods that this territory produces only in the ports or in the capital cities of the province or of the department."[50] The proposal was to restrict foreign merchants to three areas within the networks of the Peruvian internal trade: the coastal ports, the capital cities of each department (at the time: Lima, Trujillo, Tarma, Arequipa, Ayacucho, Cusco, and Puno), and the fifty-eight provincial capitals that formed part of the Peruvian territory in those days. Beyond the provincial capitals Peruvian merchants would have the monopoly in the internal trade.

Thus a conflict was evolving between foreign and national merchants for control of the Peruvian trade, in both domestic and export markets. Paul Gootenberg has studied this problem in the context of the government's commercial policies, particularly in foreign trade.[51] But what about domestic trade? We know at least that the bill of 1828 banned foreign merchants from operating beyond provincial capitals, although they could work and trade in ports and in departmental and provincial capitals. Yet, some previous questions remain: How did this domestic trade evolve? What were its areas of articulation?

The geographic and economic articulation of the internal market was largely related to mining production and trade, although internal trade and the domestic market in nineteenth-century Peru clearly exceeded the rather large dimensions of the mining economy. I will discuss the regional articulation of places, towns, provinces, and ports through this market, an articulation based mainly on trade relations, and particularly on the circulation of commodities, mostly silver ore, metal, and coins.

Until 1826 internal customs taxed domestic trade with alcabalas and other fees. But we can be sure that this legal change was not fully reflected in reality. According to Jorge Basadre, for example, "the set of rules of 6 June 1826 again canceled the internal taxation of domestic trade."[52] However, in the Peruvian National Archive are Customs Books, which recorded and taxed internal commercial transactions until 1829. These records have been studied by Magdalena Chocano and other researchers (although only until 1819).[53] One such record for Lima reveals the following geographic scheme of the circuits of internal trade:

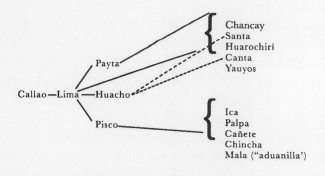

| Port | Custom | Minor Ports | Internal Customs |

Source: AGN, SHMH, Tribunal Mayor de Cuentas, OL 8, box 1, 1821.

The main port in the country, Callao, was connected to the capital city, and through it to other internal provinces: Chancay, Huarochirí, Canta, and Yauyos in the department of Lima, and Santa in the department of Ancash. The city of Lima, the capital of the nation, had a main customs house, whereas the other provinces had internal customs. Lima, and Callao, its port, were also connected with other minor ports: Payta in the far north, Huacho in the near north, and Pisco, in the near south. Pisco was the port of entry and exit for trade linking the internal provinces of Ica, Palpa, Cañete, and Chincha, with their internal customs, and Mala, which had instead an *aduanilla* (small customs office). Thus, a great commercial space was articulated through trade relations, and the government followed close behind taxing this trade in its *aduanas* (customhouses).

Peruvian historian Magdalena Chocano, who has studied trade in Cerro de Pasco at the end of the colonial period, has constructed several tables showing the regional distribution of goods that arrived in Cerro de Pasco.[54] In 1819, according to her sources, 84 percent (based on value) of the goods that entered Cerro de Pasco came from Lima. But these data, as she acknowledges, do not include goods on which *alcabala* (the colonial sale tax) was not paid: agricultural products such as potatoes, bread, corn, and wheat—all mainstays of the diet of the residents of Cerro de Pasco.[55]

The limitations of these internal customs data for tracing trade relations fully become clear when we compare the value of the mining production in Pasco in 1819 (1,523,416 pesos) with the official reports of that trade as recounted in Chocano's work (282,096.5 pesos, less than 19 percent of the total value). Presumably the remaining 81 percent went to the purchase of goods not registered by internal customs and for profits.

Still, the official records help complete the picture of the networks of regional articulation, of the multifaceted contacts that the silver trade established with other regions of the country and with other economic activities. In this sense I would like to insist on this idea of mining regions, of regional markets that through domestic trade became really economic regions in which mining was just one (albeit vital) economic component. Thus the mining trade linked mining production with agriculture, craft production, manufacturing, and livestock herding. According to the customs records that Chocano studied carefully, the mining center of Cerro de Pasco received goods from Lima, Ica, Nasca, Chancay, Cajatambo, Huaraz, Huaylas, Conchucos, Huamalíes, Huánuco, Tarma, Huancayo, Jauja, Huamanga, Huancavelica, Cusco, Andahuaylas, Puno, Salta, and Lambayeque. Hence goods came to Cerro de Pasco from nearby places like Tarma and Cajatambo, and from very distant ones like Salta and Lambayeque.[56]

Internal customs, however, disappeared legally around 1826—in reality, slightly later, in 1829 or perhaps the 1830s.[57] From then on the national government received no more revenue from the domestic trade and transportation of commodities. Trade and transportation was liberalized, and the increasing presence of foreign merchants and merchant houses was more clearly felt.[58] In 1821, for example, the year of the declaration of Independence, a foreigner, John Begg, a merchant from Scotland, was dealing with 30,711 pesos "in current money and in piña silver" in negotiations with the Lima mint and the Count of San Juan de Lurigancho. In 1825 he was a creditor to the Peruvian government for 60,000 pesos, "to be collected in London." Instead of taking payment in cash, Mr. Begg asked the government to give him the Cerro de Pasco mines of the Spaniard Juan Vivas.[59] These mines were valued at 300,000 pesos at the end of colonial times by the British traveler Robert Proctor.[60] Begg was clearly one of the first foreign merchants dealing in commercial transactions in republican Peru.

Competition over access to minerals and markets pitted foreign and Peruvian merchants against one another in the early nineteenth century.[61] As we just saw, a plan formulated by the national congress in 1828 set up new rules restricting foreign merchants to ports, and to departmental and provincial capitals. Earlier, the Reglamento of 1826 had attempted to mediate in this rivalry between national and foreign merchants. The report to the French foreign ministry on this *reglamento* by the French consul in Lima, B. Barrère, who obviously defended the interests of the foreigners and the freedom of trade, provides an interesting view of the workings of internal trade in the first half of the nineteenth century.[62]

The French consul was interested in the ease with which imported goods could make their way into the interior, from the ports to the capitals of the various departments and provinces. The consul's report was favorable in this respect: "the 6 percent of increased duties for the products introduced in the provinces has been eliminated, [thus goods] can now be transported freely from one place to another with the voucher of the customs of origin." The catch was that "this courtesy will be accorded only to natives and citizens of Peru." The merchandise would be "taken to the interior of the country" and "when in inland places, it is necessary to separate the principal invoices in order to sell some articles or to give them on consignment; these invoices have to be handled in the same way as the first vouchers taken from the customs offices, it is necessary [for selling these goods] to have two witnesses, to mention the name of traders, and the reasons for this operation." Unfortunately, in the French consul's view, the discrimination appeared again, so "foreign merchandise that was going to be traded without the documents required in the previous article will be submitted to the confiscation rule and will belong to those who denounce their trading and hold them."[63]

Many authors have discussed free trade and protectionism in the early Peruvian republic and in the early national period in other Latin American countries.[64] I am interested here in reconstructing the forms of this trade. The testimony of the French consul in Lima shows the various routes that merchandise took upon entering the country, how it was treated differently according to its market destination. This transporting and marketing of goods required continuous bureaucratic documentation to be presented to Peruvian authorities, and its form depended on whether the commodities were sold for cash or credit, or on consignment. This is a very crucial aspect of nineteenth-century Peruvian domestic trade because it deals with the forms of payment and the circulation of money. The documentation provided by the consul in this sense is very interesting and, again, shows the clear link that existed between the mining economy and regional, national, and international trade, particularly if these import commodities were paid for in silver ore, metal, or coins.[65]

In 1830 a petition from the French merchant Santiague Le Bris, "in his name and in the name of the representatives of the foreign merchant houses of Arequipa," to the customs office of the port of Islay requests "vouchers in their names to introduce their merchandise into that city." The Ministry of Finance responded:

> Having ordered that foreigners who occupy themselves in trade with the interior provinces have to move to the largest ports of the republic, it is fair to make an exception for those living in Arequipa, because between that city and the port of Islay there is no intermediate town, and the latter lacks safe warehouses and houses to live in. Warn the prefect that, while these are being built, vouchers will be given in their [the foreign merchants'] names to transport their merchandise from Islay to the above-mentioned city that they could not take nor intern in the same way and for any reason to another place in the department, nor to put them in retail trade in Arequipa.[66]

Thus we find, again, foreign imports entering the ports, Islay in this case, handled by foreign merchants who lived in the city closest to that port, in this case Arequipa (but Lima in the case of Callao, and Trujillo in the case of Huanchaco and Salaverry), and who went on to sell these goods in the internal provinces. To import the merchandise and to initiate these activities, of course, the merchants required ships, or contracts with shipowners or merchants who worked in the sea transportation business.[67] They also needed "safe warehouses" in ports and cities,[68] houses in which to live and out of which to trade, and muleteers with whom to arrange the internal trading and transport of the merchandise (selling them directly or on consignment). The French consul in Lima in this same period complained that "the government has given orders for the construction of buildings, particularly entrepots, which it intends to lease at high prices to the merchants." Thus, once more, the Peruvian government, following protectionist sentiments, planned to profit from the activities of foreign merchants operating in the country.[69]

The documents we are working with clearly show a conflict between foreign and domestic merchants for the control over and profits from trade in the Peruvian interior. According to Paul Gootenberg in *Between Silver and Guano*, the government was also involved in this conflict, sometimes favoring foreign merchant houses in brief moments of support for free trade, but more often favoring domestic merchants.[70] Gootenberg's study focuses on government treatment of merchants (mostly in Lima), lobbying groups, and government policies concerning foreign trade.[71] I can add to his historical information some trade experiences from the provinces. In the instance described below the government was trying to protect a particular Peruvian merchant house against further foreign competition, and even against the interest of Cerro de Pasco mineowners. This case also shows conflicts of interest between officials of the central government in Lima and local authorities in Cerro de Pasco.

In March 1833 the minister of finance *("Hacienda")* wrote to the prefect of the department of Junín, where the Cerro de Pasco mines were located, mandating a new policy in which at the end of each month the officers of the local treasury *("administradores de la Tesorería Principal")* had to send "the total amount of money agreed upon to be delivered by the treasury" or, at least, "the amount they actually keep." This delivery of silver bars was to be given to the merchant house of Zeballos, Iscue, and Company, which would give credit or cash money to the central government in Lima. The profits from the sale of silver bars would remain in Zeballos, Iscue's hands. The prefect, Francisco Quirós, objected to this direct intervention in favor of a merchant house in the mining business. "The measure would be detrimental for the government and the miners," wrote the prefect, because "the revenues of the local treasury could be invested in buying plata piña" locally, which would stimulate production, and also because without an official buyer of raw silver, the merchant house *("la casa especuladora")* could buy it at much lower prices. The government then, according to Quirós, would lose revenues from the increase in mining production, and the mineowners would sell their minerals at lower prices. The sole beneficiary would be Zeballos, Iscue, and Company. Evidently this monopolistic concession given to Zeballos, Iscue had its rationale in Lima, where the merchant house was currently financing government expenses.[72]

Francisco Quirós also had personal reasons for opposing this particular deal between the central government and Zeballos, Iscue. Apart from being the highest official in the area, representing local interests, Quirós was also a private merchant dealing with the silver trade between Cerro de Pasco and Lima, and a wealthy businessman in his own right with property in Cerro de Pasco and elsewhere in the department of Junín. Later, in 1855, he would become the minister of foreign relations, public education and social services *("Beneficencia")*.[73]

Another historical case of merchants involved in mining is that of Patricio Bell, also referred to in the documents as Patricio Gines or Ginez. This case offers additional insights into the characteristics of the trade and muleteering systems, and into the extent of monetization of exchanges in the mining sector. The information comes from the "summons issued against the foreigner Don Patricio Ginez, accused of clandestine commerce in gold and silver through the port of Huacho."[74] Ginez was a former Irish naval officer from the brigantine *Maypú* who arrived in Peru with the expedition of Lord Cochrane in 1818, at the age of twenty-four. That same year he was taken prisoner during the wars of Independence, and was jailed in the Real Felipe fortress in Callao until 1820, when Cochrane exchanged him for other prisoners. Once freed he went to Huaura, where he joined the Húsares Squadron of the Army Guard of General San Martín. Later in 1820 he settled in the port of Huacho, where he married Francisca Carrillo, a young woman from one of the most important local families. In Huacho, Ginez began to be involved in commerce.

On 19 August 1826, J. M. Egúsguiza wrote a letter to the intendant of the province of Chancay saying that Patricio Ginez was dedicated to the "clandestine extraction of silver, which in view of its quantity he at least must be commissioned to that effect by some merchant House of substantial size in this capital [Lima]." A few days later José Durán, an employee in the Huacho customs, declared that "Don Patricio Gines had brought from Cerro de Pasco some marcs of silver bullion ["plata piña"] with its respective voucher, which was presented at customs," but, he added, "he does not know if they were sold in Lima, placed under embargo, or were spent in this port [Huacho]." Finally Manuel Sosa, Huacho's civic lieutenant, stated that Ginez's "trade [established his career] selling grape liquor *(aguardiente de uva)*, wholesale or retail, and some foreign cloth."[75] From these accounts we can infer several conclusions. To begin with, Ginez was part of a triangular trade back and forth between Huacho, Lima, and Cerro de Pasco:

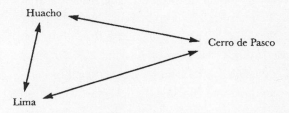

Ginez was probably commissioned by a trading house in Lima, and worked out of Huacho selling grape liquor wholesale or retail, and importing cloth; in exchange for these goods he received silver bullion *(plata piña)*. On 4 October 1826, according to the testimony of José María Pagador, customs administrator for the port of Huacho, Ginez brought to Huacho, with Lima as a final destination, "three bars of silver with a total weight of 592 marcs 2 ounces under a free voucher from the administration of Cerro de Pasco no. 31, dated 8 June." It is also recorded that Ginez owned two houses in Huacho, a schooner that in 1823 was leased to the patriot government for 700 pesos per month, and that he had "given credit for a number of pesos to various people."[76] Through the legal system, Ginez defended himself from the allegations that he had smuggled contraband, and he explained his commercial activities. He had been first a merchant of foreign and domestic textiles, and later a wholesale and retail merchant of grape liquor, which he brought by sea from the port of Pisco. He did not take money to Pisco, but instead sold on credit and payment orders *(libramientos)*. He indeed had made several trips to Cerro de Pasco:

> The first one was taking azogues to sell but I could not sell them so I had to give them on credit to the miners and financiers *[habilitadores]* of those mineral areas. The second trip I made, I was taking European goods, but again I could not sell them so I went to the city of Ayacucho, after the Junín action, and from there to the city of Ica, where I could trade the rest of the clothing I took, buying brandy, which I took to Pizco [*sic*], and from there by sea to the place of my residence. [The third trip] I took *tocuyos* [raw domestic textiles] . . . and I was basically occupied in collecting the payment for the azogues I had left before on credit.

Finally, he took the silver to Huacho and, although he wanted to send it to Lima, he sold it for stamped silver bars *(plata sellada)* and silver coins to another trader and muleteer, José Jaramillo, who came from Lima and was going to the province of Huaylas, in Ancash, another major mining area. Jaramillo bought the silver at 7 pesos 7 reales for each marc, because Ginez "was almost out of cash and it was difficult for him to take the silver to Lima to sell it at the price it deserved. The buyer [Jaramillo] was a trader and muleteer from Lima who constantly made trips to the highlands with goods from Europe, wax candles and indigo in large amounts."[77] Jaramillo later took the silver to Lima and sold it, making a profit of 4 reales per marc of silver, a total of 296 pesos in just one trade operation that lasted only a few months.

Beyond the Lima–Huacho–Cerro de Pasco triangle mentioned earlier, we now see that there are seven places touched by this circulation of goods and silver, Lima and Cerro de Pasco the most important among them:

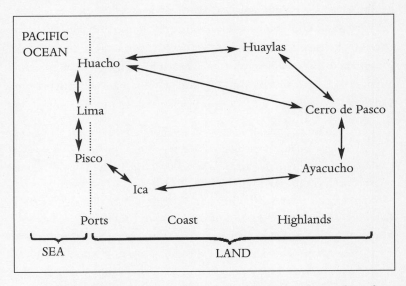

The Huacho-Lima-Pisco communication network was based on sea transportation, using the schooner Ginez owned. Between Huacho, Huaylas, Cerro de Pasco, Ayacucho, Ica, and Pisco the only means of transportation was the muleteering system. The only place in which money was exchanged was Huacho, where Ginez traded his silver bars for money with the other merchant-muleteer, Jaramillo, who in turn exchanged the bars for money in Lima. The rest of the exchanges were on credit or orders of payment (libramientos), as when Ginez bought grape liquor in Pisco, traded clothing for grape liquor in Ica or gave azogue, clothing, or grape liquor on credit *(al fiado)*.

Finally, Patricio Ginez also exchanged goods for silver bullion, plata piña, in Cerro de Pasco. In the process, Ginez forwarded merchandise on credit in Huacho. It is true that behind all this circulation of goods was the mining wealth of Cerro de Pasco and Huaylas, but when did this wealth really appear as money in these exchanges?[78] Only, as we have seen, in two areas: in Lima, the commercial and administrative capital of the country, where Jaramillo sold the silver bars for money; and in Huacho, the port and home base of Ginez, the merchant-muleteer, where Jaramillo bought his silver bars with cash. Everywhere else (Cerro de Pasco, Huaylas, Ayacucho, Ica, etc.) the commercial transactions entailed solely the exchange of goods for other goods: wheat for clothing, grape liquor for silver bars, azogue for plata piña, and so forth. Or, if it was not a goods for goods exchange, the payment was in the form of libramientos or al fiado— credit.

In short, this was a debt economy, an economy of credit, without the actual circulation of money, of coins, as a form of payment.[79] The Ginez example shows once more "the slow rotation of capital" and money,[80] the lack of velocity in the real circulation of money, of silver coins, in the regional internal trade. The transportation and commercialization systems were among the bottlenecks in the development of the mining, and thus in the economy in general, in nineteenth-century Peru.

In fact the activities of muleteers and merchants, viewed up close, forces us to redefine what I have previously called mining regions into rather broad economic regions, where silver was eventually exchanged with other goods (European merchandise, grape liquor, wheat, clothing, etc.) in a large market and pool of goods that did not always entail the exchange of money. Instead, mining products were exchanged with the products of agriculture, pasturing, and manufacturing. The principal beneficiaries of these only partially monetary transactions were a social segment of intermediaries, the merchants; the secondary beneficiaries were muleteers. These transactions were a big part of the commercial dynamic of regional markets in early-nineteenth-century Peru.

The Ginez case also illustrates the irregular topography of the country and how muleteering and trade were adapted to these rugged conditions, using as means of transportation the technology and animals available in the areas where the merchandise needed to be transported. The Peruvian coast is flat, except for minor slopes, but it is also narrow. The first slopes of the Andes appear, on average, only thirty kilometers from the sea (although this distance is greater in areas such as Piura on the far northern coast). In this narrow strip of deserts and gentle valleys, carts could be pulled by horses, oxen, or mules. In the steep ravines and canyons of the highlands, the wheel was virtually useless. In the more than two thousand years of pre-Columbian Andean civilization, the wheel was not discovered. Transportation was accomplished by llamas and alpacas, the native animals of the region.[81] Mules, brought by the Spaniards at the beginning of the sixteenth century, adapted quickly to this environment.[82] At the end of the eighteenth century a large market for mules (a mule fair) took shape in the Mantaro valley, south of Cerro de Pasco, at the Tucle hacienda.[83] And in 1874 the subprefect of Jauja, also in the Mantaro valley, estimated at 20,000 the number of mules raised in the province, where "they serve daily for the *arrieraje.*"[84]

As we have seen, the proceedings against Patricio Ginez give us much information about the nature of internal trade. Three areas are evident in Ginez's networks of exchange, each with its distinctive means of transport of goods: the sea, with its coastal trading by ship; the coast, with its wagons and carts pulled by horses and mules; and the sierras, with their mules and llamas used in arrería trading. Specific data on mules are scattered. However, in 1826, the year of Ginez trial, the British consul in Lima stated that "the only modality of transportation for the goods that go to the interior of the country are the mules." He sought the total amount of goods traded domestically in Peru at that time, as well as the amounts of U.S. and European commodities. "Although I could not check the annual amount of European and North American goods transported to the different districts, the amount transported to the provinces of Huaylas and Tarma from Lima was estimated at 8,000 mule loads *[cargas de mula]* of 10 to 12 arrobas each, or a weight of 2,200,000 pounds."[85]

Much later, in 1889, Julio C. Avila and Ulises Bonilla discussed the role of beasts of burden in the transportation systems of the mines of Parac and Colquipallana: "the only carriage *[arriage]* that is easy to obtain here are llamas, for mules and burros one has to go to neighboring places."[86] In 1846, J. M. McCulloch offered a panorama of the Peruvian livestock industry of European origins at the time:

> The ordinary horses and the mules could be bought between 45 and 50 dollars each. Piura is well known for its excellent ranching of the latter, and many mules are taken from there to go to Trujillo, Lima, etc. where they reach sometimes the price of 250 dollars each. The same province is also famous for its goats. Many hogs are also raised in Peru, they are considered ready for the market at 10 to 16 months of age when they are sold for 6 or 9 dollars each if they are of good pedigree.[87]

Mules then, and llamas, were the basic beasts of burden, the "trucks" of nineteenth-century Peru. They carried the goods that circulated throughout the country's various regional markets.[88] These mules were raised on specialized farms, and llamas in communities of peasant herders *(comunidades de pastores)* that existed (and still exist) throughout the Andean highlands, as nineteenth-century historical and twentieth-century anthropological evidence shows.[89]

Leaving aside the goats and hogs of McCulloch's observation, let us return for a moment to the mules. A substantial number came from Piura, where they were raised in large numbers. At 250 pesos each, five times their price in the area of production and twenty-five times the price of a hog, mules were highly valued as a means of transportation and other work, whereas hogs were only a consumer good. The Piura mules had been central to trade since colonial times, when they were used for the transportation of goods from Cuenca in Ecuador to Lima, following the *tierra firme* route. According to Argentine historian Silvia Palomeque, at the end of the eighteenth century and the beginning of the nineteenth, "in Piura the mules are hired for the second section of the trade route [Piura-Lima], taking the road that passes through Lambayeque and the Guaura and Santa Rivers."[90]

The price of 250 pesos per mule in Trujillo and Lima was also between five and ten times higher than the price of mules a century earlier: in 1754 they were sold for 25 pesos in Piura, 30 pesos in Cajamarca, 40 pesos in Lambayeque, Pataz, and Cajamarquilla, and 50 pesos in Tarma.[91] In this last place, in the central sierras of Peru and south of Cerro de Pasco, 3,000 mules were allocated for transportation, which represented a cost of 150,000 pesos, or 75,000 in real terms, because a portion was paid in *"ropas de la tierra," "jerga,"* and *"bayeta"* (crude domestic cloth).[92] These 150,000 pesos, the total value of mules in the region, represented 37.5 percent of the total value of goods to be distributed in the region, according to the colonial *repartimientos* (the forced distribution of goods). This percentage shows again the strategic importance of mules for the transportation of goods in the Peruvian domestic market.

The higher prices of mules in the nineteenth century reflected their higher demand in expanding regional markets, where these beasts of burden were indispensable for the efficient transportation and distribution of commercial goods. In 1846 Carlos Renardo Pflucker, a mineowner in Morococha, noted in his *Exposición que presenta al Supremo Gobierno* that he hired a company of muleteers from Piura, paying them 10,000 pesos, particularly during the summer months, when the muleteers of the Morococha area were occupied in the transport of snow and ice for sale in Lima.[93] In this example from mid-century, Piura is once again the supply area for mules and muleteers.

Patricio Ginez, our merchant from Huacho, used mules and horse carts for the transportation of goods in the coastal areas, between Pisco and Ica, in Lima, and in Huacho itself. But above all he used the sea, transporting goods on his schooner between the minor ports of Pisco and Huacho, and the major port of Callao. Between Lima and Callao the obvious means of transportation was wagons and carts. In the highlands however, between Huacho, Cerro de Pasco, and Ayacucho, and later the route down to Ica, the basic means of transportation was again mules and, in some cases, llamas.

The observations of the British consul in 1826 and the references of Ginez allow us to estimate quantities and values of goods transported that year. Unfortunately we do not know the exact commercial value of this trade particularly between Lima, Huaylas, and Tarma, but we can estimate it at more than 3 million pesos, considering an estimated value of 1.5 pesos per pound of commodities bought and sold in this exchange network. We have to remember that among these commodities precious metals comprised a minor proportion, the larger part being goods for daily consumption. On the other hand, the Ginez trade between June 1825 and April 1826 included three silver bars with a total weight of some 260 pounds which earned him 4,392 pesos, or 5300 pesos annually.[94] Ginez, however, obtained these silver bars after selling domestic and foreign textiles, azogue, wheat and grape liquor, among other goods. Thus Patricio Ginez, or Patricio Bell, was just a medium-sized merchant in Huacho, working in connection with Cerro de Pasco, the great mining center of the central sierra, and Lima, the Peruvian capital. He was a medium-sized merchant in a sea of goods.

It is important to note too that Patricio Ginez was both an independent merchant and muleteer, and a commissioner of a Lima merchant house. As an independent trader, he traveled with his mule and llama teams to the places mentioned above (Cerro de Pasco,

Ayacucho, Ica, Pisco, etc.), probably hiring other muleteers to help him with his business. As a representative of a Lima merchant house, that much larger commercial enterprise (the name of which is not mentioned in the proceedings) was really in control of a trade network in which Ginez was just a small part. The Lima merchant house operated in a trading network that touched, again, several places in the central region of the country (Lima, Huacho, Cerro de Pasco, Ayacucho, Ica, Pisco, etc.).

Trade, then, and particularly trade in silver, was a lucrative economic activity in early-nineteenth-century Peru. It connected various areas of the country, creating and developing regional markets. It brought various economic sectors into contact, exchanging their productive outputs. These exchanges relied on various modes of transport (muleteering, llamas, carriages and carts, ships). Transportation by mule and llama prevailed in the internal market. However, suddenly a noisy machine appeared on the Peruvian landscape that would change the routes, characteristics, and dimensions of this domestic market, a machine that exhaled a curious dark smoke. The railroad had entered Peruvian history.

Chapter 5

Railroads and Muleteering

Coexistence and Conflict

UNTIL THE RAILROAD appeared all internal transportation in the Peruvian domestic market was through the muleteering system. The railroad dramatically changed the pace and nature of the mining economy, although its full effects, particularly in the mining centers of Casapalca, Morococha, Yauli, and, more important, Cerro de Pasco, became clear only in the twentieth century. Plans to build railways began in the 1820s, although construction did not begin in earnest until the 1850s. It then boomed in the late 1860s and early 1870s. This boom was quickly followed by a financial and economic crisis that eventually paralyzed railroad development.

Muleteering, then, never disappeared but continued to play a role in the mines, and in the Peruvian transportation system as a whole. Indeed this role still persists today. Thus, railroad and muleteering coexisted in a long transitional period that started in the 1870s and acquired new dimensions with the building of roads and highways, and the presence of the automobile and the truck, already in the twentieth century. Railroads and mules kept separate routes and areas of operation, and fought against each other for the control of some of the benefits of the nineteenth-century mining economy.

Why did the railroad enter the Peruvian social economy so slowly? Why did it not immediately change the dynamics of mining? Why did the railroad not supplant muleteering? Was the impact of the railroad stronger in mining or in some agricultural sectors? Finally, did the railroad contribute to the building of a nationally integrated economy with a powerful domestic market, or did it instead consolidate export economies?

A Newcomer to Arriería: The Railroad from Beginning to Boom

Railroads were at the center of modernization of the means of transportation in late nineteenth-century Peru. In 1826 three businessmen living in Peru (Francisco Quirós, Guillermo Cochran, and José Andrés Fletcher) signed a contract with the government to construct what would have been the first steam engine railroad in Latin America, a project that was never realized.[1] In 1834 and 1835 contracts were signed for railroad projects that never materialized.[2] In 1848 the building of railroads in Peru began and thereafter passed through four periods. The first, between 1848 and about 1860, centered on the construction of the first Peruvian railway between Lima and Callao, twelve kilometers long and costing 550,000 pesos. This work was done by the Peruvian businessmen Pedro González de Candamo and José Vicente Oyague "y Hermanos."

A second railroad, connecting the inland city Tacna with its port, Arica, was sixty-three kilometers long and cost 2 million pesos. The businessman José Hegan backed its construction between 1851 and 1856. The third railroad built during this period was the Lima-Chorrillos line. Fifteen kilometers long, it connected Lima with a minor port and suburb, Chorrillos. Chorrillos was a fishermen's cove and, like Barranco and Miraflores, became in the second half of the nineteenth century a coastal resort where Lima's elite built beach houses, lured by the recent "spiritual discovery of the sea."[3] The work was done between 1856 and 1858, again by Pedro González de Candamo, at a cost of 350,000 pesos.[4]

It is between this last year, 1858, and about 1868 that the second period in railroad building began, promoted especially by the General Law Governing Railroads, upon which the future President Manuel Pardo commented with great optimism: "Who could deny that the railroads are today the missionaries of civilization?"[5] This excessive statement has motivated some historians to argue that Manuel Pardo, as the ideologue of triumphant liberalism—the liberalism of guano and the emergent Civil Party—was the leading representative of the "new Peruvian hegemonic faction" or of the "landed-commercial bourgeoisie." [6] In any case the building of railroads gathered speed and by 1874 eleven private lines were in existence (see table 5.1).

In the first period of railroad building, the 1850s, the three lines constructed totaled only ninety kilometers. By the end of the second period the amount of track had been multiplied: 439 kilometers in use, which should have reached 590. Note too that although the table Martinet compiled extended to 1874, the lines of Pisagua–Sal de Obispo, Pimentel-Chiclayo, and Cerro de Pasco had yet to be completed.[7] The Cerro de Pasco line was directly related to mining production because it linked the mines within the city's limits with the refining mills on its outskirts.

Some of the railroads built in this period were small urban passenger lines, such as the one linking Lima with La Magdalena. Others played a more clearly economic role, linking areas of production to ports, as in the Salinas de Huacho–Playa Chica or Chancay-Palpa lines. But there were also the long and important rail lines that fulfilled key economic functions, such as the Iquique–Nueva Noria and Peña or Pisagua–Sal de Obispo lines, that were directly related to the extraction and transportation of nitrates in the far southern coast.[8]

TABLE 5.1
Railroads Built or in Use by Private Companies in Peru, 1874

Railroad Line	Length (km.) In Use	Total*
Lima-Callao	12	12
Lima-Chorrillos	15	15
Iquique–Nueva Noria and Peña	113	113
Pisagua–Sal de Obispo	80	175
Eten-Chiclayo and Ferreñafe	85	85
Pimentel-Chiclayo	24	72
Tacna-Arica	63	63
Cerro de Pasco	11	19
Salinas de Huacho–Playa Chica	10	10
Lima–La Magdalena	6	6
Chancay-Palpa	20	20
Total	439	590

*Includes track to be built.
Source: J. B. H. Martinet, *La agricultura en el Perú* (Lima, 1877; reprint, Lima: Centro Peruano de Historia Económica, 1977), 97. I have included in the table the railroad lines mentioned before (Lima-Callao, Tacna-Arica, Lima-Chorrillos), because they were also in Martinet's list.

These railroads were, as we can see in map 5.1, coastal lines, extending over the relatively flat Peruvian littoral and scarcely penetrating inland, into the Andes. They were built, furthermore, over flat lands, which eased their construction and transit and also lowered their costs. As we can also see from the map, the railroad lines covered only a small part of the national territory. There was only one line in the north of the country; it revolved around the activities of the city of Chiclayo and the valley of the Chancay River, where sugar production in haciendas was growing. A few lines were concentrated on the area around Lima, from Huacho to Chorrillos. This proliferation of railroad lines around the capital was another indication of an ever-present centralism in the organization of the nation. Everything revolved around Lima, even for this early period of railroad construction. Finally, there were a few lines in the south, from Tacna to Peña.

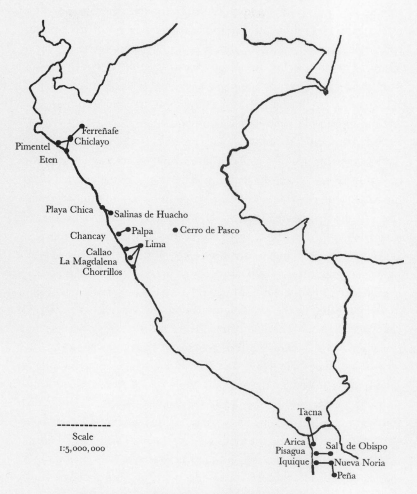

Map 5.1. Railroads Built by Private Companies, 1850–1870

The Iquique and Pisagua railroads, in the old Peruvian province of Tarapacá, certainly had been built in response to the prosperous economic dynamic of the region, which had seen a revival in silver mining since the 1850s, based on investments of capital that came from the nitrate business.[9] Tarapacá had eighty silver mines in 1878 and 103 in 1879.[10] The need for railroads to transport the minerals and mining supplies was therefore obvious.

In the Lima area the building of railroads was also very active. The Salinas de Huacho–Playa Chica line evidently had as its goal the transport of salt blocks to the port. The Chancay-Palpa line linked the valley with the port taking agro-pastoral goods to the Lima market, as well as cotton exports, which were multiplying.[11] The lines of Magdalena and Chorrillos were also railroads that, apart from passengers, brought staples to the Lima market from orchards, vegetable plots, small farms, and haciendas located just outside the capital city. Recall that Lima in 1876 had 120,994 inhabitants and eight urban markets located in the five neighborhoods *(cuarteles)* of the city. It also had 2,412,320 Castilian varas (some 168 hectares) of vegetable plots *(huertas)* in the southern section of the city, inside the city walls.[12] Feeding this metropolis, in other words, required supplying it with provisions from the surrounding areas. Martinet, also the author of a report on the "scarcity of foodstuffs in Lima,"[13] confirmed this fact describing the two new railroads connecting Lima with its surrounding provinces (the Chancay-Ancón and Lima-Pisco): "Both offer the advantage of ending in Lima, facilitating the provision of the markets of this city and those of Callao with a quantity of foodstuffs for daily consumption."[14]

In summary, the 1850s and 1860s saw the beginnings of railroad construction, bringing these new machines to still relatively restricted areas of the country. Trains created modern, fast transportation of large quantities. They connected economically important urban centers with productive areas, such as the developing cotton and sugar exports in the Chancay valley and Chiclayo, the growing nitrate ex-

ports in the province of Tarapacá, and mining in Cerro de Pasco and
Tarapacá. But this was just the beginning. The real boom in railroad
construction would not begin until 1868.

Foreign Capital and the Boom in Railroad Building

It is in the third period of railroad building, from the administration
of President José Balta (1868–72) to the financial crisis of 1876, that
the railroad reached its apogee in Peru. By that time the young minister of finance Nicolás de Piérola had given exclusive control of the
guano trade to the French merchant and financier Auguste Dreyfus.[15] Thus began what Basadre has called "the era of great public
works." It is now the Peruvian government that would promote railroad construction, engaging in negotiations with an American businessman and adventurer, the "Yankee Pizarro," Henry Meiggs.[16]

In this flurry of new construction amazing engineering feats were
accomplished. For example, the Central Railway, ascending toward La Oroya, in only 142 kilometers reached an altitude of 3,724
meters at Chicla, crossing sixty-five tunnels and sixty-one bridges.
Construction of many of the new lines proceeded at an accelerated
pace. The Mollendo to Arequipa line, for instance, was completed
in only thirteen months, five months ahead of schedule. This massive new construction required an investment of 132,052,357 soles
(£20,529,772),[17] and 10,000 to 12,000 workers labored at any one
time under Henry Meiggs's control.[18] Closely following Martinet's
account, table 5.2 reflects this extraordinary boom in railroad construction.

In the 1870s railroad construction continued its fivefold expansion.[19] No other result could be expected from such a massive investment of capital. During the railroad boom of the 1870s two core
networks penetrated the Andes: the Mollendo-Arequipa, Arequipa-
Puno, and Juliaca-Cusco lines would be joined into a single line,

TABLE 5.2
Peruvian Railroad Lines Built by the Government or by Companies with Government Participation, 1876

Line	Ownership	Length in use (km.)	Total Length	Cost (soles)	Cost per kilometer
Callao–Lima–La Oroya	SO	146	219	21,666,860	98,935
Mollendo–Arequipa	SO	180	180	12,000,000	66,667
Arequipa–Puno	SO	370	370	25,120,997	67,895
Juliaca–Cusco	SO	–	354	23,959,144	67,681
Chimbote–Huaraz–Recuay	SO	102	265	24,000,000	90,566
Pacasmayo–La Magdalena	SO	146	146	5,850,000	40,068
Salaverry–Trujillo–Ramales	SO	–	88.5	3,234,756	36,551
Paita–Piura	SO	–	100	1,945,600	19,456
Ilo–Moquegua	SO	101	101	5,025,000	49,752
Lima–Ancón–Chancay	P	66	66	2,600,000	39,394
Pisco–Ica	P	74	74	1,450,000	19,595
Lima–Pisco	P	–	260	5,200,000	20,000
Total		1,185	2,223.5	132,052,357	59,389

SO = fully state owned; P = with private participation.
According to contract; in cash.

Sources: Martinet, *Agricultura en el Perú*, 99. A similar table with some differences in length and costs is in Esteves, *Apuntes para la historia*, 143. Finally, there is another set of data with costs in pounds sterling in Spenser St. John (British consul in Lima), "Informe General sobre el Perú," 1878, in Bonilla, ed., *Gran Bretaña y el Perú*, 1:173–99, esp. 197.

forming the Southern Railway *(el Ferrocarril del Sur)*. Likewise the Callao-Lima-La Oroya line would become the Central Railway (see map 5.2). Thus a comprehensive system of transportation by rail started to spread all over the country.

By this time the Southern Railway was not only the longest line in the nation (904 km.), but also the busiest. In 1877, when Martinet was writing his report, the section from the port of Mollendo to Arequipa was already finished, as was the section from Arequipa to Puno, on the shores of Lake Titicaca. Shortly afterward el Ferrocarril del Sur would reach Santa Rosa, a town between Juliaca and Cusco. The regional impact of this railroad on the economy and society of "Arequipa and the Andean south" has already been explored by historian Alberto Flores Galindo.[20]

Other railroads of significant length in this third period of railroad building were the Central Railroad (219 km.) and the Lima-Pisco line (260 km.), although the latter never proceeded past the planning stage. After the start of its construction had been delayed, the 1876 financial crisis hit and the Peruvian government suspended payment of its external debt, provoking a monetary crisis. Nor was the Lima-Pisco line revived in the fourth period of railroad building, after the war with Chile. The Central Railroad, on the contrary, reached the little town of Chicla, 146 kilometers from Lima, where construction stopped until the end of the War of the Pacific. After the war's end it reached La Oroya in 1893, Cerro de Pasco in 1904, Huancayo in 1909, and Huancavelica in 1924.[21]

The highland railroads that penetrated the Andes were understandably more expensive to construct than the coastal lines. A coastal railroad like the Pacasmayo-La Magdalena, Salaverry-Trujillo, or Paita-Piura never required more than 6 million soles in capital. In contrast, a highland line that extended deep into the Andes cost from 12 to 25 million soles. The average cost per kilometer for construction of a coastal line was 32,116 soles; for the average highland line it was more than double, 78,348 soles.

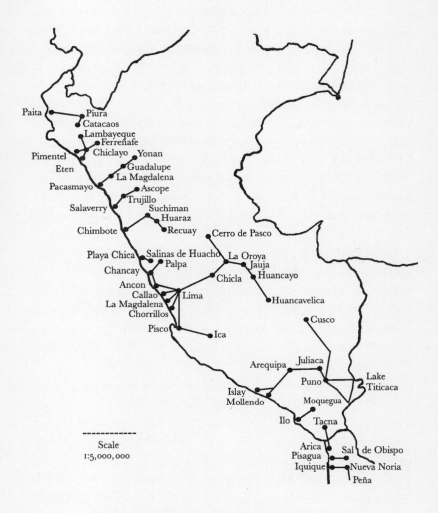

Map 5.2. Peruvian Railroad Lines, 1850–1900

In 1875 and 1876 the Peruvian government suffered an acute financial crisis, directly related to the breakdown of the guano industry, that forced it to stop the work on the railroads. A fourth period of railroad building would not begin until 1884, after the war with Chile, once a national reconstruction began. In 1889 the foreign debt crisis was resolved with the signing of the Grace Contract, which stipulated that the Peruvian government turn over control of the public railroads to the bondholders of the foreign debt for sixty-six years. A foreign enterprise, the Peruvian Corporation, was organized to manage the railroads.[22] During the era of the Peruvian Corporation some lines were finished and others were lengthened. For example, the Central Railroad, as I said before, reached Cerro de Pasco, Huancayo, and, in 1924, Huancavelica. The Southern Railroad reached Cusco, the Convención valley, and finally in 1951, the city of Quillabamba.[23]

In the mid-twentieth century there were 3,263 kilometers of rail line in all Peru. The lines of the Peruvian Corporation were the most extensive (1,734 km.), followed by those of the Cerro de Pasco Corporation (333 km.), which built a specially designed section between La Oroya and Cerro de Pasco to transport the minerals taken from its mines. The government at this time controlled only 20 percent of all the nation's lines.[24]

The influence of large foreign companies like the Peruvian Corporation and the Cerro de Pasco Copper Corporation continued to increase and eventually became an overwhelming presence within the Peruvian railway network and the mining industry. As the national economy became more capitalistic, large foreign companies that had entered the Peruvian market through the railroads extended their activities into mining, as did the Peruvian Corporation when, through the Grace House, it formed the Empresa Socavonera del Cerro de Pasco on 26 October 1900.[25] This metamorphosis in such foreign companies also reflected the transition from silver to copper mining. The latter required true industrial complexes with refineries, and smelting and concentration plants. These industrial complexes were built at the end of the century, starting in the 1890s, in Tamboraque, Casapalca, Tinyahuarco, Cerro de Pasco and, finally, La Oroya.[26] The railroads, and railroad building as a business, were hence the vanguard of a more industrial, capital-based, mining economy in which the mass production and processing of industrial metals, especially copper, took place.

The Structure of the Railroad System: Development of the National Market or of Export Economies?

During the railroad construction fever that began in the Balta administration, railroad lines grew in number and length.[27] I have obtained statistics that show the importance of these lines in terms of the yearly shipping of freight and passengers between 1890 and 1900, during which time railroad service was being consolidated and in full operation. These statistics provide a picture of the structure of the railroad system in late nineteenth-century Peru.

Consider first the transportation of people. There were three levels in the Peruvian railroad system of the late nineteenth century: (1) The railways around Lima (Lima-Callao, Lima-Chorrillos, Callao-Lima-Chicla [the Central Railroad]) carried over half a million passengers each year. These were the largest railroad lines in the country. (2) The railways of Eten-Chiclayo-Lambayeque, Salaverry-Trujillo-Ascope, and Mollendo-Arequipa-Puno–Santa Rosa (the Southern Railroad) were part of a second level. Each of these lines transported more than one hundred thousand passengers per year. (3) The third level was formed by the seven other railroad lines that each transported fewer than 72,000 passengers per year.

A better indicator of the importance of the railroad lines in terms of moving men and women around the country was the ratio between the number of passengers transported and the number of kilometers

TABLE 5.3
Transportation of Passengers and Freight by Peruvian Railroads,
1890–1900

Railroad Line	Passengers	Freight (tons)
Paita-Piura	31,043 (10)	36,431 (5)
Piura-Catacaos	34,652 (9)	1,755 (13)
Eten-Chiclayo-Lambayeque	119,007 (6)	16,160 (7)
Pimentel-Chiclayo-Lambayeque	14,827 (12)	6,500 (11)
Pacasmayo-Guadalupe-Yonán	58,441 (8)	21,094 (6)
Salaverry-Trujillo-Ascope	149,884 (5)	50,983 (4)
Chimbote-Suchimán	2,905 (13)	3,761 (12)
Lima-Ancón	71,601 (7)	13,493 (8)
Callao-Lima-Chicla	678,631 (2)	103,867 (2)
Lima-Callao	601,190 (3)	159,512 (1)
Lima-Chorrillos	909,181 (1)	9,555 (10)
Pisco-Ica	16,603 (11)	11,194 (9)
Mollendo-Arequipa-Puno–Santa Rosa	204,155 (4)	55,411 (3)
Totals	2,892,120	489,716

Yearly averages.
Numbers in parentheses indicate rank.
Source: Heraclio Bonilla, "El impacto de los ferrocarriles: Algunas proposiciones," Historia y Cultura (Revista del Museo Nacional de Historia, Lima, 1972), 93–120, tables 7–8. Bonilla's sources are the Anales de la obra pública (Lima, 1899), 619–27; and documents of the Peruvian Corporation kept at University College, London. Bonilla notices, however, some differences in the data of both sources that he tries to resolve, see esp. 99.

these lines covered. Again, by this index the semi-urban lines around Lima were the biggest: Lima-Chorrillos with an index of 60,612 passengers per kilometer, Lima-Callao with an index of 50,099 passengers per kilometer. The first connected Lima with neighboring towns like Miraflores and Barranco. But the second line also fulfilled an important economic function, linking the capital with its port. Apart from these two, the railroad lines of the Central Railroad (3,098 pass./km.), Salaverry-Trujillo-Ascope (1,693 pass./km.) and Eten-Chiclayo-Lambayeque (1,400 pass./km.) followed in order of importance.

In the transportation of goods, the railroad system had a different hierarchy. One line clearly surpassed all the others: the Lima-Callao railroad, which linked the capital city with its main port, and transported some 160,000 tons of goods annually. The Central Railroad carried more than 100,000 tons of goods per year. The Southern Railroad, the Salaverry-Trujillo-Ascope line, and the Paita-Piura line all transported between 36,000 and 55,000 tons per year. Four lines on the coast (Pacasmayo-Guadalupe, Eten-Chiclayo, Lima-Ancón, and Pisco-Ica) connected ports with valleys producing agrarian goods for export, such as cotton, sugar, or rice,[28] and carried between 11,000 and 22,000 tons per year. A final set of lines transported under 10,000 tons of goods per year (Piura-Catacaos, Pimentel-Chiclayo-Lambayeque, Chimbote-Suchimán, and Lima-Chorrillos).

This railroad structure is central to the formation and consolidation of a modern export sector at the end of the century. The line that carried the largest amount of freight was the one linking the capital city with its port, Callao, whereas the second one was the railroad that penetrated the central sierras developing a fast and modern exit route for the mining production of the area (Cerro de Pasco, Yauli, Casapalca, Morococha, etc.), as well as for the agricultural and pastoral goods produced in the Mantaro valley and its highlands *(punas).*[29] Many of the other lines (Salaverry-Trujillo-Ascope, Paita-Piura, Pacasmayo-Guadalupe, Eten-Chiclayo, Pisco-Ica) that transported between 11,000 and 55,000 tons of goods per year were clearly connecting ports with areas that produced agricultural goods recently in high demand on the international market. Modern plantations of sugar and cotton, for example, developed in the Chira, Piura, Lambayeque, Saña, Jequetepeque, Chicama, Moche, Virú, and Chao valleys (to focus just on Peru's northern coast), and their production was largely oriented toward the international market.[30]

Rail lines were also constructed within coastal haciendas. Whereas one kind of line (often standard gauge, 1.43 meters wide) connected the main entrepot in the valley to the port, whether Guadalupe to

Pacasmayo, Chiclayo to Eten, or Ica to Pisco, other rail lines were fully within haciendas.[31] The latter in general were narrow-gauge lines and therefore less costly to build. The same phenomena took place in Andean mining centers, in places like Cerro de Pasco and Puno, where mineowners built their own wagon lines within their mines. These short, narrow lines connected private mines with the main railroad lines that linked mining centers with cities and ports (as did the Central Railroad with the mining center of Cerro de Pasco, and the Southern Railroad with Puno). Landowners and mineowners built their own private lines within their units of production (haciendas or socavones), whereas the national railroad network tied their economic enterprises with an increasingly developed international market that depended heavily on buying large amounts of raw materials.

As shown in table 5.3 the transportation of passengers and freight was not always of equal importance for a given line. The Paita-Piura line, for example, was the fifth largest in bulk cargo but the tenth in transporting people, while the Lima-Chorrillos line was almost exclusively a passenger line.

The table proves, however, that the Peruvian railroad system, whether in terms of passengers or freight, only served to consolidate the growing centrality of Lima. Between 1890 and 1900, an average of 2,260,600 passengers a year took trains in or out of Lima. This amounted to 78 percent of all passenger train travel throughout the country. A similar centralism is visible in freight: 286,427 tons, or 58 percent of all goods transported by rail, entered or exited Lima.

The capital city grew demographically, economically, and in terms of the market and transportation systems built around it. At the same time, a modern export economy developed, one that linked agricultural, pastoral, and mining areas with ports and international trade markets. The growth of the Peruvian railway system in this context, and its impact on the national economy, is shown in map 5.2.

Peru since the 1850s was becoming a country of railroad tracks and locomotives (compare, for example, maps 5.1 and 5.2). Modernity appeared to have arrived. However, in spite of this immense growth, much territory was not connected to the railroad network. Early-twentieth-century data show that Peru had 3,263 kilometers of railroads, but this was a ratio of only 0.33 kilometer of track for each 100 square kilometers, placing Peru seventeenth among the twenty largest Latin American countries.[32] The 2,000 miles of Peruvian railroads at this time were dwarfed by the 15,000 miles in the Mexican network.[33]

A comparison of maps 5.1 and 5.2 also shows that four new lines were constructed in the north and a large line (for Peru) was built in the central areas, as well as the Chancay-Ancón-Lima, and Pisco-Ica lines on the near northern and near southern coasts, respectively. In the south a very long line was also established, the Southern Railroad, in addition to a smaller line between Ilo and Moquegua. However, a large portion of the southern lines, indeed of the entire Peruvian system, would be lost in the war with Chile between 1879 and 1884. The Pisagua–Sal de Obispo and Iquique-Peña lines—commercial, mining and nitrate routes—as well as the Tacna-Arica line, became part of Chilean national territory after the war, when the provinces of Tacna, Arica, and Tarapacá, which included Pisagua and Iquique, were ceded (the first two only provisionally) to Chile after signing the Peace Treaty of Ancón in 1883.

Close scrutiny of map 5.2 reveals several rail lines that spread from the Peruvian littoral, from the ports and the sea, to extend into agricultural valleys on the flat lands of the Peruvian coast and eventually reach the foothills of the Andes. Only two lines, the Central and Southern Railroads, crossed the Andes, 12,000 to 15,000 feet high, to connect with mines, haciendas, and peasant villages in the Andean highlands.

Many areas in the Andes, however, remained unconnected to railroads: all the northern sierra, for example, from Conchucos to Cajamarca, and also the central and southern highlands, from Huancavelica to Cusco. Even today the departments of Ayacucho and Apurímac lack contact with this modern means of transportation. In those large areas where the locomotive and the railroad tracks never appeared, mule and llama trains continued to be the only known means of transportation. This continues to be true today in spite of the building of more recent highways and roads.[34] The eastern part of the Andes that reaches the Amazon tropical forest was also untouched by the railroad. This central jungle of Peru contained areas like Chanchamayo where the production of sugar, sugar-cane liquor, coca, and coffee started to grow.[35] In these areas muleteering was connected with the Central Railroad, which ran from Jauja to Huancayo, in a complementary axis from Jauja to Tarma to Chanchamayo.

To some extent this process of railroad building integrated the country. At the same time, it caused even more dislocation, by creating within the same national territory economic spaces that had different rhythms of economic development. Some regions now had better access to modern means of transportation, and hence to markets and capital. Whole regions in which the railroad did not penetrate, however, such as Ayacucho with its "glorious colonial past" and its thirty-three colossal and architecturally grandiose churches, would experience a backwardness and a sluggish development whose consequences are still felt.[36]

Gordon Appleby, referring to Puno, from which he derives his dendritic model of regional marketing systems remarks that "when inter-regional transportation improves without a similar improvement in the intra-regional transportation system, which was historically the real significance of railroad building, the differences in the transportation systems create local commercial centers at the side of the railroad lines with their only goal to concentrate the small amounts of raw materials to be sent to the ports."[37] Therefore, even in regions where the impact of the railroad transportation network was felt, inequities in social and economic development persisted.

Raw materials were extracted from areas without rail transportation to satisfy a commercial demand elsewhere.

In the second half of the nineteenth century the impact of the railroad was stronger on the Peruvian coast, where a boom occurred in the cultivation of export cash crops, particularly sugar and cotton. The central sierra was also changed by the railroad, although to a lesser degree than the coast. The railroad intensified the commercial relationships of the highlands with the large market of Lima and, through the port of Callao, with the world economy. This is the region in which the influence of railroad transportation was stronger on mining, although it was also the area in which the steel horse more sharply confronted the competition of muleteering. Finally, the southern highlands would also experience a boom in the export of wool, both sheep and alpaca, in which railroad development was a major factor.[38]

There is no doubt that the railroad played a significant role in causing this moment of commercial and export expansion.[39] It was not, however, the best way to consolidate national integration or to build a strong domestic market. Rather, as some authors have observed before, the railroad was a speedy mechanism to make the Peruvian economy more dependent on the international market. The railroad was a major factor in transforming the Peruvian domestic economy into a series of export economies.

But how did this wave of railroad construction, say from the 1870s on, affect the mining economy? What happened to all the commercial circuits created and developed by muleteering? Did muleteering itself suddenly disappear?

Railroads and the Persistence of Muleteering in the
Nineteenth-Century Mining Economy

Jean Martinet wrote in 1877: "The working of the mines has almost always been the goal of the construction of a network of railroad

156

lines; agricultural production, however, has not been forgotten, and today it obtains more advantages from transportation by rail."[40]

Recalling the mining regions mentioned in the previous chapter, we can see in map 5.2 that only some railroad lines were oriented toward mining zones. Certainly this is the case of the railroad to Cerro de Pasco, which arrived in Chicla in 1875 and in La Oroya in 1893. By this second date the operations of this railroad line had already connected several mining centers in the Lima sierras, such as Casapalca, Morococha, and Yauli. Not until 1904 would it reach Cerro de Pasco.

The Chimbote-Suchimán line, destined for Huaraz and Recuay, would establish contact among several mining centers in the provinces of Ancash, in the near northern sierras of Peru. These mining centers in the department of Ancash were part of the 226 mines that existed in 1878, and of the 448 mines being worked in 1887.[41] The Pacasmayo-Yonán and Salaverry-Ascope lines reached the foothills of the northern Andes, where, at the highest altitudes, the mines of the department of La Libertad and, above all, of Cajamarca, were located. Ica, connected to a port by the Pisco-Ica line, was also a mining crossroads where the flow of mining goods reached the city from Cauza. The Southern Railroad similarly established contact between mining centers, such as Puno and Lampa, although its first economic goal, as I have already said, was the transportation of sheep and alpaca wool.

Martinet's observation, made more than a century ago, is accurate: the impact of railroads on mining was significant, particularly in the last third of the nineteenth century, although it was even greater on agriculture and livestock herding. The clearest example of a railroad built in, and for, a mining center, was the Cerro de Pasco line, which extended throughout this area. Its construction began at the end of the 1860s. Eleven kilometers had been laid in 1877, falling short of the goal of nineteen kilometers. The government resolution that ordered its construction was approved on 26 September 1867 and ratified on 28 March 1868. The company working on its construction was Enrique Orbegozo Wyman y Compañía, and the railroad was to extend from Cerro de Pasco to the town of Pasco, passing through Quillacocha (*sic,* probably Quiulacocha), Sacrafamilia, and Tinashuasco (probably Tinyahuarco).[42] This line transported ores from the mines of Cerro de Pasco to the refining mills (*ingenios*) in valley heads (*quebradas*) such as Quiulacocha, Tambillo, and Sacrafamilia.

However, neither the line's dimensions nor its economic impact on mining production were great. Tracks were laid, and wagons to carry ore were bought, as was a steam locomotive. But the costs of the whole venture (construction, management, and administration) were too high, and the railroad company had to charge high fees for its services. Thus, wagon trains pulled by oxen, horses, or mules, using the old routes and roads, remained competitive.[43] Old means of transporting ores to the ingenios, and metals from there to areas of trade, using even llama trains, then were very much alive. And they operated at the same sites where the railroad tracks were being built. Thus, in an 1887 report on the Cerro de Pasco *"asiento minero"* (mining center), I. C. Bueno stated that in those years "the price for transporting metals from the mine yard *[la cancha de la mina]* to the refining mills *[haciendas de beneficio],* located at different distances from Cerro de Pasco, when it is done on the back of beasts *[a lomo de bestia],* varies from 15 to 40 centavos for each load of one quintal. When it is done by the railroad, it depends on the prices that the railroad company charges."[44] Small-scale mining production perhaps did not yet need the speed and capacity of the railroad. Transportation "a lomo de bestia" could still well satisfy local market conditions. The choice of mules versus trains depended, according to the Bueno report, on the fees the railroad charged for transporting minerals. If they were high, both mineowner and refiner might prefer the old arriería system. The keys were the amount of mineral to be transported and its type. The smaller scale of silver mining, for example, supposed the production of a highly priced precious metal, whereas copper mining involved the extraction and processing of larger quantities of ore. Silver mining, then, could be accommodated by the arriería system, whereas copper required the railroad. By 1887 large-scale mining in Cerro de Pasco was not yet a reality.

If the acceptance of the steam locomotive and a railroad company was still not total in the mining center of Cerro de Pasco in 1887, wagons on iron tracks pulled by mules, horses, and even oxen were better received. Long rail lines not only traveled between mines and ingenios, but they also were used increasingly within mines to extract ore and waste from the bottom of mining shafts to the surface. In 1892 these rails and wagons were in full use in the extractive process in Huarochirí and Yauli. One observer used the word *barrenos* to describe the workers who "in the mine shafts lined with tracks *[los socavones enrielados]* push the wagons to take the mining waste and ore out of the mines."[45] Although the steam locomotive was still not in universal use, railroads clearly saved time and money in the transportation of heavy loads, as the British historian Rory Miller has previously remarked in reference to the Peruvian central sierra.[46] But it is also true that the miners, producers, and transporters continued to follow traditional practices, resisting the wholesale use of the railroad while adopting some of its elements over which they had greater control in terms of costs and operation. This simultaneous resistance and adaptation to a new technological change in mining production, previously observed by Carlos Contreras, created a relatively long period of transition and coexistence before the new conditions of transportation, commercialization, and economic production introduced by the railroad came to prevail.[47]

To understand this transition, roughly between 1870 and 1900, it is helpful to keep in mind that the railroad was a technology foreign to the Peruvian economy. At the time of its introduction, muleteering was perfectly compatible with the level of activity of the mining economy. The railroad was rather the result of technological achievements of more developed countries. To operate, the railroad needed steam engines and locomotives, modern devices to drive and maintain the machinery, trained technicians to operate and repair this equipment, steel-frame wagons for both passengers and freight, large amounts of coal, and so on. The investment required to import these elements, or to reproduce them locally, had to be high.[48]

Rory Miller argues that there was not enough domestic capital from mining to pay the costs of railroad building. Even in Cerro de Pasco the mineowners did not have enough capital to finance the construction and operation of a railroad linking Chicla and the mining city. According to Miller this explains why it took so long for the railroad to reach the mining area and when it did it was constructed by the Cerro de Pasco Copper Corporation, a foreign company with large amounts of capital at its disposal.[49] The railroads were thus an extension of the export economy, heavily dependent on foreign capital, and not the result of the internal development of domestic resources within the Peruvian economy.

And if investment costs were high, the foreign investors also wanted quick and abundant returns. Therefore they immediately pushed the local miners to pay the investment costs, charging high fees for the use of railroad services, as the Bueno report of 1887 acknowledges. Railroads, as an imported technology, thus implied heavy costs for the domestic economy.

Under these conditions the Peruvian government was forced to subsidize part of these investment costs. For example, in 1874 the minister of government in his *Memoir* of 1874 contended that construction of the railroad that connected the mines and the refining mills in the area of Cerro de Pasco should continue because "the needed materials have already [been imported]."[50] Similarly, about the Arequipa railroad Basadre states that "two million soles of the twelve million needed for the work should have to be deposited by the government in any of the merchant houses dealing with guano to buy materials abroad *[en el extranjero]* for its construction."[51] Thus, in the first case the government was involved in the importation of materials needed to build railroads; whereas in the second case the government paid some of the investment costs using economic resources coming from guano exports in order to import railroad equipment. In both cases, to achieve its goal the government had to deal with private merchant houses, mostly foreign.

For local miners, then, it seemed logical to continue to use the old arriería system, or even llama trains, because their monetary costs were much lower. In 1893, an engineering student, Carlos Y. Lisson, described operating conditions in Parac, a mining area "two hours from San Mateo," in the Lima sierras. He mentioned that to operate the railroad there and connect it to the Central Railway in Chicla required buying coal at forty soles per ton, while llameros charged twenty centavos for carrying one quintal of mining goods as "freight on the back of llamas" *(flete a lomo de llama),* and the cost of taking a beast from Parac to San Mateo was just two soles.[52] Does this mean that the railroad was a bad investment? To answer this question, one would have to consider various macro- and micro-economic effects, as well as the opportunity costs. During the railroad construction boom 132 million soles were invested to improve and accelerate the transportation systems of the country. Might this capital have been effectively deployed in other projects or initiatives?

The answer to this second question is probably yes. But railroad building was part of a complex reality that went beyond economic calculations; it was also part of an international wave of ideological perceptions of development and modernization, from which Peru was unlikely to insulate itself.[53] The presence of the railroad started to disrupt the traditional ways of transporting and trading minerals, consumer goods, and supplies to the mining centers, but these traditional ways resisted the assault. However, the Peruvian domestic market began to change. Railroads meant large capital investments, new foreign companies, high-speed transportation, bulk cargo, and so on. Were they not beneficial for the mining economy? Was it possible to do without them?

In order to survive, muleteering tried to lower transportation costs. A period of resistance, conflict, and adaptation ensued in which muleteering competed hard against the incorporation of the new foreign and costly technology. As a result, this incorporation would not be complete until the next century, particularly in the central sierra and Cerro de Pasco.

However, the situation was somewhat different in a region closer to where the Central Railroad was being built, the Lima sierras, with the mining centers of Casapalca, Yauli, and Morococha. The impact of the railroad was stronger there, changing the characteristics of the regional market and particularly the role that the town of Chicla played in it. With its introduction the railroad also transformed trade relations and the regional economic circuits.

In 1875 the Central Railroad was extended to the small town of Chicla, 142 kilometers from Lima. Chicla was only thirty kilometers, a day's walk, from the mining centers of Casapalca, Yauli, and Morococha. The railroad brought its cargo of goods and passengers, who from Chicla on had to use mules and llamas to reach the mining centers. Between January and March 1889, Michel Fort, a university student who would become a mining engineer, went to research the area:

> Leaving Lima via the trans-Andean railroad . . . one reaches the town of Chicla, 3,724 meters above sea level, and currently the final stop of the above-mentioned railroad. Trade is well developed there because it is the place where commercial merchandise and minerals are stored to be later transported to the coast. It is also where the mines are provided with the articles they need most. Concerning the purchase of the latter, each mine has an agent, either in Chicla or in Lima, who is in charge of filling the various orders of the administrators. Usually these are materials for working the mines, such as oil, dynamite, ma-

chinery, etc., or, perhaps, necessary supplies and staples that cannot be obtained in the highlands or that are purchased more advantageously than in the interior economy. These supplies and staples are goods such as rice, sugar, wheat, biscuits, etc.

And later, in reference to the goods that the "interior economy" transported to Lima, Fort argues:

> The articles that come from the interior are usually carried by Indians who own them or are in charge of selling them on a commission. Those in this second category are less common because it is almost impossible to trust people who, because of their customs or lack of education, cannot think for themselves. These articles are potatoes, corn, cheese, etc.[54]

Chicla was thus transformed into a permanent commercial fair. Every arrival of the railroad, every departure, changed the quiet life of what Tschudi had called many years before "a poor Indian village."[55] Now the railroad arrived full of staples and daily consumables, such as rice, sugar, wheat, biscuits, and mining inputs (oil, dynamite, machinery), and departed full of minerals and other goods. The Andean peasants were drawn into the new commercial dynamism, taking to market their excess production (potatoes, corn, cheese, etc.). The mines had agents in Lima or Chicla (or both) who were in charge of selling minerals or buying or filling purchase orders for staples and mining inputs.

From Chicla toward the highlands' interior, however, toward the mining centers of Casapalca, Yauli, Morococha, toward Cerro de Pasco and, still farther, to the central sierra (Jauja, Tarma, Huancayo), trade and transportation continued to be by mule train, burro, and, especially in the case of the Indian peasant, llama. Another engineering student, Baldomero Aspíllaga, observed in 1889 that "muleteering is easy to obtain in Chicla." He also discussed the fees that muleteers charged and compared them with those charged by the railroad companies. According to his report, the fees that muleteers with llamas charged from the mining centers of Huarochirí to Chicla were 15 centavos when the transportation was going downhill and 20 centavos when it was uphill. Burros were more expensive: 80 centavos downhill, one sol uphill; and mule trains cost 1.80 soles downhill, 2 soles uphill. So, "transportation is most of the time done by llamas while it is very rare when it is done by burros or mules, unless very heavy loads must be transported." Finally, he states, "this mineral center [Huarochirí] only uses llamas, which makes them very frequently part of the capital of the mining companies."[56]

Similarly, in Cerro de Pasco, llama drivers represented one of the highest costs in at least one large, relatively modern mining enterprise. The mining company (La Negociación Minera) of Erasmo C. Fernandini between August 1883 and June 1884, calculated its "operations with llameros," at 42.8 percent of its costs, or 176,367.10 soles out of 412,401.10.[57] In fact, llamas were used extensively in the Lima sierras, in Huarochirí, and in Cerro de Pasco, even after the railroad reached the town of Chicla. Nevertheless, the railroad transformed Chicla from a "poor Indian village" into a regional commercial center whose role was now to articulate the rail transportation system, from Lima to Chicla, with the muleteering system, from Chicla to various mining centers and other areas of agrarian production.

Outside the central sierra, in areas where the impact of the railroad was even weaker, we obviously find that the role of muleteering in transporting mineral or other commercial goods was greater. In 1890 in Hualgayoc, Cajamarca, for example, where the rail lines from Pacasmayo and Salaverry had not yet reached the mining centers, José Antonio Araoz emphatically argued that it was impossible to develop mining further with the existing system of roads and mules:

The roads that link Hualgayoc with the coast, with Cajamarca, Bambamarca, Pilancones, and other neighboring places, are all, as is usual in the highlands of Peru, rugged and tortuous; but they pose no danger. During the dry season they offer easy access and transit, but during the rainy season they become very difficult to pass, whether because of the mud that forms near swamps that obstruct them, as on the roads to Yanacancha and Pilancones, or because of the rivers and torrents that reach extraordinary dimensions, to the point of not allowing any traffic at all.[58]

Araoz constructed a table showing the distances between Hualgayoc and the most important towns, places, and mining centers of the region. Between Hualgayoc and the district of Bambamarca, for example, it was 4 leagues; to Yanacancha, where there were soft-coal mines that supplied fuel for the refining and processing of silver ores in Hualgayoc, the distance along these regional roads was 5 leagues. To Tallamaj, an anthracite mine, 6 leagues; to Trascorgue, the location of amalgamation and smelting mills, 2 leagues; to Pilancones, where there were more amalgamation mills and yards *(ingenios y patios)*, 2 leagues; and to Cajamarca, the capital of the department, 12 leagues.[59] All these distances were traveled by mules and llamas. The arrieros who led them were designing the circuits and routes of this regional mining market. With their constant movement they integrated the various aspects of mining (production, commercialization, transportation) into a single process. For Cajamarca, at the end of the nineteenth century, the railroad was still only a dream.

Araoz's report emphasizes again and again the difficult conditions under which minerals were transported in one direction, and mining inputs and agricultural supplies in the other. These conditions, according to his testimony, prevented any relationship between the mining center of Hualgayoc and the international economy through the export of minerals.[60] These problems of transportation and communication also prevented the technological renovation of the mining enterprises in Hualgayoc, making difficult the building of a concentration plant in the area:

Not long ago this important business of exporting valuable and carefully selected minerals started. But in order to reach its full potential, this business would require, understandably, the building of concentration plants *[oficinas de concentración]*. These are very costly, almost unfeasible, because the plants would require machine tools and equipment that, because of their size and weight, would be impossible to transport on the back of a mule.[61]

Thus, these two transportation and marketing systems—muleteering, the heir of a long colonial past, and the railroad, a relatively recent product of the European Industrial Revolution—coexisted in the Peruvian mining social and economic conditions of the late nineteenth century. Even in the early twentieth century, the new technology could not displace the traditional, less efficient one within the capitalist system.

Although the two systems occasionally complemented each other, the conflict between muleteering and the railroad hindered the economic reorientation of mining production toward the international market, toward export growth, and toward the consolidation of large mining enterprises and the proliferation of smelting, amalgamation, and concentration plants. Such a reorientation would have imposed (and later did impose) on the national economy of Peru the new social and economic rules of capitalist market efficiency, time savings, and transportation speed brought by the presence of the railroad. But the refusal of muleteering to die at the end of the nineteenth century is a symptom of the resistance of peasant ways of life to disappear when confronted by a new world of business and heavy industry.[62]

An explanation for the endurance of competition from muleteering, as I have said before, was the high fees that the Peruvian Corporation charged for the transportation of freight. This was the result of the near monopoly that the corporation tried to create in the national railway system, and certainly with the Central Railroad to Chicla. In 1890 still "a third of the minerals that came out of Cerro de Pasco reached Callao on the shoulders of mules."[63] Although the Peruvian Corporation used several freight tariffs according to the type and quality of ores transported and the different sectional railroad lines traversed, transportation costs were still very high to producers, whether owners of mines or ingenios. In 1892 Ismael Bueno compared the costs and benefits of both systems in Cerro de Pasco:

> [Shipments of] high-yielding ores that exceed 100 marcs are exported in spite of the very high freight fees, especially the fees from Chicla to Callao. Before, they charged 11 centavos for every 100 kilograms per kilometer, independent of the quality and grade of the mineral. Today those minerals that exceed 100 marcs are charged 30 centavos. It is for this reason that it is more expedient to refine the less yielding or second-class minerals in the same place, and not to export them.[64]

Bueno's report includes a table showing the costs of taking 100 marcs of silver ore to the international port of El Callao by rail. His calculations compare muleteering and railroad costs:

> The cost to take one box [*un cajón*] of 100 marcs of silver from Cerro de Pasco to Callao is the following:

Sacks	18.00
Freight from Cerro to Chicla, at	
4 soles per load [*S. 4 la carga*]	80.00
Commission	10.00
Railroad costs from Chicla to Callao	115.07
Total	S[oles] 223.07 [65]

Note that the railroad from Chicla to Callao cost more than the muleteering charge from Cerro de Pasco to Chicla, although the distances were about equal. Bueno thus clearly supports the muleteer system: "It is more convenient to send the metals directly to Lima via the Obrajillo route, where 9 soles are charged for each load of 12 *arrobas*, or one box, of mineral. Adding other expenses, it would cost 180 soles plus a small fee from Lima to Callao."[66]

The monopoly that the Peruvian Corporation held over the railroads contributed to the high cost of transport by rail. Bueno states that due to a sudden management decision transportation prices almost tripled (from 11 to 30 centavos to carry each 100 kg. of silver one km.). With muleteering, on the other hand, it was always possible to lower costs, or to shift some monetary costs to natural costs, maintaining or even raising, monetary benefits. Francisco R. del Castillo, who traveled the regions of Huarochirí and Yauli in 1891, observed that the use of llamas and even people, mostly men, to carry minerals, was a way to lower monetary costs. The efficiency of transportation, however, decreased, preventing technological modernization. The work of llamas and people, especially Indian peasants, was an economic tool cheaper than the railroad, from the standpoint of the mineowner. The reproduction expenses of llamas and Indian peasants were lower, and they were more useful for the transportation and commercialization of silver.[67]

Llamas, in monetary terms, cost little, especially if they were raised on the property of the mineowner or Indian peasants. They reproduced according to a natural cycle, and if there were costs they were not monetized, but were part of the family costs in a peasant economy. The grasses *(ichu)* on which the animals fed were also often free if they belonged to peasant community land or a hacienda. Indian peasant porters were paid little and sometimes not even in money but in goods produced locally by haciendas and peasant farms. Large capital or monetary investments could be avoided in the muleteering system. Finally, it was always possible to translate monetary costs into natural costs, or to demand more sacrifices from the peasant economy or from peasant resources.

On the use of llamas, del Castillo wrote:

> The beast of burden commonly used is the llama, which, as is well known, is an animal suitable to these places and is also a reliable investment. This animal needs no care of any kind, nor any special fodder. It takes care of itself, and from llamas we obtain everything.
>
> For nourishment they take plants that grow in the neighboring hills. For transport, only males are used, and for reproduction and light work, females. They can carry up to fifty kilograms for about six

leagues. They start to work when they are two years old and work until they are eight.

Their meat is an Indian staple, their skin is used for the shoes *[ojo-tas]* of the same Indians, and their wool for the weaving of textiles and clothing, and even to make *huascas,* or ropes to tie their loads.

To do their work llamas are divided in groups of thirty that labor fifteen consecutive days and then rest almost the same amount. A muleteer using llamas on a working journey is called a *chacanía*.[68]

Del Castillo also described the 300 workers, the Indian peasant operarios, who labored in the mines and who came "from the neighboring towns and even from faraway ones, such as the town of Jauja": "The transportation of heavy machine parts is extremely difficult and can only be done by humans. Thus, to bring from the neighboring hills stones for the *vastie* [a piece of equipment for the ingenio], many of them work several days, with the help of a lever."[69]

Hence mules, llamas, and cheap human labor transformed monetary costs into "natural" costs, and could keep competing with the railroad. The persistence of inexpensive traditional modes of transport is attested by the claim of Julio Avila and Ulises Bonilla in 1889 that transport for the mines of Parac and Colquipallana, in the central highlands, "could still be done using llamas, mules and burros." And, they continue, "llamas carry one quintal of ore from the mine to Parac, in the same way that they do from Aruri to Parac, with the freight cost of one real. For the heavy ore or machinery parts, burros and mules could be used, where the limit a mule carries is ten arrobas and the freight cost is in line with the usual rates of the area."[70]

Similarly, Julio Morales, commenting on mining in the province of Huarochirí in 1892, argues that

> transportation is accomplished using llamas, burros, or mules, but llamas are more common. Each llama carries one quintal and the costs are the following for each quintal of mineral transported: from the Eliza mine to Rayo 5 centavos, and to Chicla 30 centavos, from Milagros to each of these places is 5 more centavos, from Eliza to Casapalca 10

centavos. In Aguas Calientes from the headquarters [administrative office building] to Casapalca 5 centavos and to Chicla 10 centavos. From Jirca to the house 5 centavos.

. . . these prices are not fixed by custom and are not completely stable. Heavy loads are sent by mules and their prices are fixed by custom. Transportation, then, is cheap and, if we consider the great number of muleteers ready to carry freight, the result is that it is easy and convenient.[71]

In Morales's words, muleteering was cheap, easy, and convenient. The transport of large quantities of mineral and the need for speedy delivery did not yet change the preference of many toward the railroad.

Morales's report also shows the relationships that were established between the mineowner and muleteers or llameros, relationships mentioned above concerning the Negociación Fernandini in Cerro de Pasco for the years 1883 and 1884,[72] ten years before Morales's fieldwork. According to his report, "each hacienda has a group of llamas under the supervision of an arriero, called a chacanía. These chacanías, with the money they obtain for their transport services, pay for the llamas little by little until they own them."[73]

The mineowners, then, also had haciendas, *fundos,* or *parcelas* (small or medium-size landholdings) where they could raise llamas. They also had *peones,* farmworkers or tenants who also served as muleteers for the transport of minerals. When this peón, herder *(pastor de hacienda),* or farmworker did his job as transporter, he was paid, although not necessarily in money but with increasing ownership of the animals. After some time the llamero would possess the llamas, although the mine- and landowners also had more animals on their haciendas, the offspring of the original herd. Thus, mining, animal herding, and muleteering had many points of contact, interpenetrating each other in their functions as a "traditional" economic activity. Traditional mining, animal herding, and muleteering were also, to some extent, an ecological unit, because they all used the natural habitat of llamas, the Peruvian puna, as a base for their economic operations.[74] In this context, the railroad was surely an external element.

However, when the muleteer or llamero owned the animals and the mineowner did not, the relationships between them changed. Morales stated that "the arrieros who do the transportation with their own llamas demanded to be paid in cash, and they are called maquipureros."[75] In this case, cash circulated, not debt relations or patron-client arrangements. The muleteer or llamero was in this case an independent entrepreneur, although he could belong to a peasant community or to a peasant village and thus be part of a collective or family-based economic strategy. In any case, he was paid for his services in cash.

All these social, economic, and cultural relationships were destined to disappear with the introduction of the railroad, although we could obviously expect that they were determined not to. According to the data provided by the British consul Alfred St. John between 1896 and 1897, mineral exports from the central sierra reached 1,800,000 soles in bars and plata piña from Cerro de Pasco, 1,200,000 soles in bars and plata piña from Casapalca, and 750,000 soles in raw minerals containing silver (minerales argentíferos) transported by the Central Railroad (el Ferrocarril Central).[76] A larger amount of refined minerals, with more value added (bars, plata piña), were exported than the crude mineral ores containing silver. And this larger amount of refined minerals was not transported by the railroad.

This information shows the still low impact of the railroad on the mining economy, only four years before the beginning of the new century. The function of the railroad was the transport of raw ores of lower value, instead of silver metals processed locally. The railroad was not still lowering the overall costs of production of the mining economy in the central highlands. It was still more profitable to refine raw ores with high silver yields on the spot. Perhaps this was due, again, to the high costs of transportation and, therefore, to the high fees that the railroad charged for its services.

The data of Alfred St. John, however, show that now the railroad was carrying large amounts of ores and metals. Between 1892 and 1897, then, the railroad had gained some ground. We do, however, have to take into account that St. John was focusing on the central railroad and not on the alternative system, muleteering.

The question of whether to process silver ores locally was also the concern of Ismael Bueno in 1892. At that time he argued that because of a recent increase in rail fees it was more beneficial and profitable to locally process minerals with more than 100 marcs of silver law. It was preferable to send other minerals as bulk via the railroad.[77] If that was the case in 1892, it continued to be so in 1897.

In any case, railroads, mules, and llamas still coexisted as mineral carriers at the end of the century, competing against each other. Yet sometimes modern and traditional modes of transportation complemented each other. In Cajamarca, José Antonio Araoz recommended the elimination of the muleteering system, but also highlighted ways in which the two complemented each other.[78] Mule and llama routes were changed to adapt to, and connect with, rail lines. In Cerro de Pasco and the central sierra, even as the railroad lines grew and connected strategic places (Chicla, La Oroya, Cerro de Pasco), the muleteers foresaw these new conditions and tried to adapt (establishing new connections between mines, lowering costs to compete with the railroad, etc.).

The competition between the two means of transportation and marketing, however, was not limited to costs. Because of its ability to carry large, heavy loads and its speed, the railroad was a more efficient means of transportation to and from the mines. The Callao-Lima-Chicla route took only one day; the mule trains took seven or eight.[79] Thus, the railroad also sped up commercial transactions, hastened the delivery of goods, minerals, or mining supplies, and brought markets closer to production centers. Time increasingly became a crucial element for economic enterprises, augmenting also the velocity of capital movement and investment opportunities. With the new century the railroad was more firmly entrenched within the mining economy, particularly in the central highlands of Peru.

A final factor leading to the eventual rise of the railroad as the preferred mode of mineral transport was the sheer volume of minerals to be moved. When, as a result of accelerating investment dynamics, the volume of mining output grew, and hence what Miller calls the kilogram/kilometer unit of measure,[80] it was a propitious shift for railroad transportation. The yearly average of 103,867 tons (nearly 104 million kg.) of commodities transported by the Central Railroad between 1890 and 1900 could not have been handled by mule and llama trains alone. It would have taken 100,000 mules, each making ten trips, to carry this amount of goods in one year, or one million mules making one trip. Such a scenario is unimaginable. The railroad was raising the amount and speed of the exchange of goods, changing the context in which regional markets operated in early-twentieth-century Peru. The 1890s was then the turning point for rail transportation in mining.

The railroad also meant, as this chapter has shown, a reorientation of trade and transportation routes, managed now by railroad companies. The muleteering system did not disappear, but its subordination to the railroad was neatly established by the end of this transition period of coexistence and conflict. It was evident that a new era was beginning for mining production. But this dawn still had to wait until copper mining transcended the social and economic rules set by the silver-mining industry. The new rules of copper mining meant larger capital investments, larger foreign companies operating in the country, and the production of an industrial good on a larger scale than ever before.

Chapter 6

Conclusions

Mining and National Development

THIS STUDY BEGAN by confronting the question of knowledge in these times of postmodernist disbelief. In the introduction I quoted from Aristotle: "All men [and women] naturally have an impulse to get knowledge." Perhaps, by knowing many voices from the past, we will be able to see the present and even the future more clearly. Perhaps we will fulfill the promise of the Latin American proverb, the past is the prologue of the future.

Antonio Gramsci conceived of historical knowledge as the applied science, the praxis, of the future. He was extremely concerned with the idea of action, of the agency of human beings in the historical process, particularly workers, but most important, of peasants in the Mezzogiorno of southern Italy. In clear contrast with other versions of "real socialism" being implemented in the 1920s (Lenin, Stalin, even Bukharin and Togliatti), Gramsci believed that we all have some understanding of historical events, of the evolution of human actions. He believed that all men and women are philosophers. Hence his idea of "common sense," a notion related directly to questions of ideology and ideological control, of hegemony and "counterhegemonic" projects of development.[1]

History, in a Gramscian perspective, is the effort to comprehend the cumulative human experience that interconnects past, present, and future. It is a "science of action" for the present and future, based on the analysis of the past. In this book I have tried to impart a better understanding of Peruvian historical reality in the nineteenth and, I suggest, twentieth century by examining the bewitchment of silver in nineteenth-century Peruvian mining. What then did we learn about mining in nineteenth-century Peru?

Mining was key for the Peruvian economy in the nineteenth century and continued to be so in the twentieth. But the development of an economic sector is the result of the presence of various social, economic, cultural, and political factors, of individual and collective choices and decisions. The growth of mining in nineteenth-century Peru as an export economy rather than an internal engine for domestic development was only one of several possible paths for Peruvian mining and economic development.[2] The choice of one over the others was the result of a historical process I have tried to grasp here. The muleteering system and the railroad collided in the second half of the century, as did industrial mining of copper versus silver. The latter was a more traditional sector, although better linked to the internal market.

However, these alternatives were internal as well as external. The Peruvian economy could not isolate itself from the course of an international economy that was marching down a path from the domestic capitalism of several European countries, the United States, and perhaps Japan, toward world capitalism. The crisis in nineteenth-century Peruvian silver mining, as we saw in chapter 2, could not be seen apart from the drop, and later collapse, of the international price of silver.[3]

This book started also confronting the question of knowledge due to the increasing specialization of disciplines in the social sciences, the atomization of intellectual inquiry, and the recent influence of some postmodernist schools of thought and historical analysis that border on skepticism or, perhaps, nihilism, à la Nietzsche. Umberto

Eco wrote recently about the tendency to overinterpretation that some postmodern currents have in historical and, in general, social, literary, and cultural analysis. According to him, although these modern currents of thought are called postmodern, they look instead "very pre-antique."[4] Eco's cautionary recommendation seems even more alarming since he himself was the one who instigated the rise of postmodernism and the debate about textual meaning with the publication of his book *Opera aperta* in 1962.[5]

In my effort to track the mining sector in nineteenth-century Peru, my goal has been empirical rather than interpretive. I have tried to present the reader with large pieces of new information, of new primary data that has to be handled and understood in some way. Thus, this study has not only "discovered" new facts but also has tried to recapture the multiple voices of the past that could be heard and read through numbers. A kind of music emerges from this study, a kind of poetics. All through the century socavones were dug by Indian workers under the command of caporales and mayordomos; mules, carrying heavy loads of silver, were driven and later traded by arrieros on the various routes and markets of the country; since the 1870s locomotives, making "an infernal noise," arrived in Callao, Lima, Morococha, Cerro de Pasco, Juliaca; finally, indigenous peasants looked at them with amazed eyes, covering their ears when these "exotic" machines that "run so fast" pulled into the stations. Thus a postmodern reading could have been done from this historical research. I chose not to do it.

I also looked for meaning behind the facts or descriptions of social conditions and daily life, which most of the time consisted of a kind of economic behavior. This study could then be called traditional, and I would be proud of that label. In opposition to some recent currents in social science history, I claim that there is "down there," underneath all the descriptions and facts, a social and economic reality, a social economy, which can be studied, from which we could learn, and which might offer us a prognosis about the future, à la Gramsci. Allow me now to briefly summarize the main conclusions of this work.

Mining in Nineteenth-Century Peru: Foreign Capital, Silver Mining, and the Internal Market

Throughout the nineteenth century mining was a key productive sector for the Peruvian economy and society. It generated constant revenue and a commercial and social dynamic. It was certainly not a modern capitalist sector; but mining was always a dynamic industry, although experiencing, as any economic activity, cycles of growth and decline, however preindustrial those cycles were.[6] Mining was not an "annihilated industry," an economy in crisis with closed mines, or a nonexistent activity, as some authors have claimed. It was the opposite: an enduring economic sector that produced several million pesos worth of minerals per year, extracted from close to 2,000 mines on average, and employed at any given time between 5,000 and 9,000 mine workers.

However, this key sector of the national economy did not experience a surge in production until the end of the century. Only in the 1890s, with the displacement of silver by copper, can we speak of an economic transition to a more modern and capital-intensive industry. In the 1890s foreign capital arrived in the country in large amounts and the scale of production changed, as well as the goal of mining production. Now the logic of the industry was to extract very large amounts of a raw ore to be transformed into an industrial metal (copper), rather than small amounts of a precious metal (silver) to be used as currency or for luxury consumption.

We have also seen that whereas the value of silver remained stable or decreased in the second half of the nineteenth century, particularly after the price crisis of the 1870s, exports of copper (ore or metal) were already increasing. Patterns in the evolution of Peruvian mining in the second half of the nineteenth century, then, forecasted the turning point that took place at the end of the century.

At the end of the century the organization and consolidation of the Cerro de Pasco Copper Corporation was a major event in the transformation, in scale and nature, of mining production, a transformation conditioned by the presence of large amounts of foreign capital.[7] The Cerro de Pasco Corporation was just one example among several, although perhaps the most outstanding. For instance, the Backus and Johnston Company started its operations in Casapalca, in association with Ricardo Bentín, in the 1880s. In 1889 Backus and Johnston had 200,000 dollars gold, or 414,000 soles, worth of investments in mines and a modern refining plant displaying the latest machinery for the concentration of metals.[8] We have also seen in chapter 3 that by the 1870s there were already foreigners involved in silver mining. In fact, the three largest mineowners in Cerro de Pasco in 1878 were foreigners. The dimensions and nature of mining production changed with large investments of foreign capital at the turn of the century, but this was a slow and cumulative process that had started a few decades earlier. Small-scale mining and the production of silver still defined the industry in the late nineteenth century.

As we saw in chapter 4, over half of all mines in 1878 were silver mines, increasing to four-fifths in 1887, after the War of the Pacific. This small-scale, labor-intensive mining economy was based on thousands of mines in several regions of Peru, but particularly in the central sierras and in the mining center of Cerro de Pasco. This silver-mining economy continued to use the colonial patio system for processing ores into silver ingots (plata piña). Silver circulated from mines to mills (ingenios), to smelting plants (callanas de fundición), and finally to ports and cities, in which silver bars were minted into silver coins, the currency of the time.

In spite of being a small-scale industry that used old colonial technologies for processing ores and that depended on a forced peasant labor supply, silver generated millions of pesos per year in value.

This commercial wealth circulated within the national economy, giving impulse to the structuring of an internal market. Silver mining, however, was also a leading export sector, particularly until the late 1840s, when guano became predominant. Nevertheless silver continued to have commercial significance throughout the guano boom. In the 1870s silver eclipsed guano and regained a leading position among export sectors, competing with new agricultural businesses such as sugar and cotton.[9] Peru has always been a *país minero,* a mining country, from colonial times to the present.

Silver, however, more than guano, combined this dual capability of being an important commodity for the development of the domestic market and, at the same time, an export business. Muleteers and merchants played a key role in this articulation of domestic market and export economy (see chapters 4 and 5).

The building of a railroad system and later the arrival of foreign companies with large amounts of capital and new technologies disturbed this accommodation between silver mining and the development of an internal market. The railroad system created a transportation network that more quickly and efficiently linked the mining economy to international markets and, with many restrictions, to an "enclave economy." At the end of the century Peru was embarking on a different path of development, one that was more open, more dependent on foreign capital, capital-intensive technologies (such as the railroad), and the international market based in London, Hamburg, Paris, or New York. As Rosemary Thorp and Geoffrey Bertram put it, Peru was becoming an "open economy" or, in the words of Florencia Mallon, the period between 1895 and 1930 saw a "national attempt at modernization" with "penetration of foreign capital."[10]

Was it possible for the Peruvian mining economy in the early nineteenth century to experience an upsurge? What would have been necessary for this to take place and what obstacles lay in its path? I offer some insights and partial answers to these questions, but with trepidation. My trepidation comes in part from the fact that I have studied a single economic sector, which, though critical to Peru's exports and for the development of the internal market, was dwarfed by the agrarian sector. Agro-pastoral production, in terms of population and therefore employment, full of haciendas and peasant communities, was the largest sector in the nineteenth-century Peruvian economy. Furthermore scholars are still discussing the magnitude and nature of the GNP in nineteenth-century Peru.[11] There are of course good sectoral studies—on the agrarian economy, for example, or the better-researched guano economy.[12] But the consistent data necessary to address these questions fully are still missing.

I will therefore try to address them in a hypothetical way, based on the historical experience of the mining sector and on my review of nineteenth-century Peruvian historiography on other social and economic sectors. I will proceed to divide the "economy," the social economy, in three traditional and classical dimensions: capital, labor, and natural resources. This analytical division, however, makes for an extremely simplistic interpretive model.

Formalizing Social Reality: Capital, Technological Innovations, Labor, and Natural Resources

This study has proved that for most of the nineteenth century domestic capital formation in the mining economy was low. Miners depended on merchants for capital supplies and for the trading of mining goods and other commodities. The investments in the productive sector were low and, thus, productive capital was, in general, scarce. However, there were some outbursts of capital formation, particularly due to the guano boom. An influx of foreign capital was also influential.

One missing factor for an upsurge in nineteenth-century mining was thus the lack of productive capital, of money invested and rein-

CONCLUSIONS

vested constantly and increasingly in the mining economy. There were no formal financial markets in nineteenth-century Peru, nor banks until the 1860s; and when these banks did start to operate, their use of credit was more speculative than productive.[13] Previously, when in the 1820s and 1830s bancos de rescate (public banks established to purchase silver bars) were created as an initiative by government and mineowners, they failed due to lack of funds, in other words, due to a lack of domestic capital formation. More important, the failure of the bancos de rescate was due to the strong opposition of merchants, who mostly controlled the mining business and were not at all interested in this competitive public venture.[14]

Although capital was scarce in nineteenth-century Peru, fortunes were not. Mineowners' wealth was defined in terms of property and status, not in terms of profits, of capital. The sector in society that made the most profits from the industry and who had an idea of profit making, albeit corporatist and monopolistic profit making, were the merchants. They controlled the circulation of silver, and the supply of goods to the mining centers. They controlled the credit business, the circulation of capital, to be borrowed at a high interest rate. The merchants' business was closer to usury and personal relationships than to money lending in a competitive market. They were the ones who had the most control over the different regional markets that formed the nineteenth-century Peruvian economy. The mining sector, the productive area, on the contrary, almost always lacked capital. When there were investments, and particularly productive investments, such as in the socavón of Quiulacocha between the 1820s and the 1840s, these were of rather small dimensions and centered on using larger quantities of peasant forced labor rather than investing in technology or research.[15]

Did the guano economy not create capital, even large amounts of it? It did. But this capital was largely speculative and superficial. The nature of the guano trade, as a boom-and-bust short export cycle, similarly generated short-lived fluid capital, some of which went to luxury consumption and, therefore, provoked an increase in expensive imported goods, creating a deficit in the overall trade balance. Some of this speculative capital generated by the guano rent went also to increase the government bureaucracy, the services and public works that the government implemented in the 1860s and 1870s (e.g., the building of railroads and the purchase of military equipment, including several iron-plate navy ships acquired during the administration of President Pezet). Involved in many of these endeavors were several foreign merchant houses and businessmen, such as the House of Gibbs, W. R. Grace, Henry Meiggs, and, above all, Auguste Dreyfus. They profited extensively from these endeavors and therefore from the capital generated by guano, which then left the country.[16]

Other destinations of the capital generated by guano were the banking industry, which also boomed and busted in less than twenty years; some productive agricultural investments in sugar and cotton coastal estates; and some, of course, in mining, although not great amounts. But, above all, the rent and profits of guano, and the immense foreign debt that it produced, were invested in railroad building, those "railroads to nowhere."[17] Guano then created, in the words of Jorge Basadre, a "fictitious prosperity" (una prosperidad falaz).[18] Capital, in that context, had a bubblelike quality.

The debate on the role of guano in the nineteenth-century Peruvian economy, of course, is not settled. Whereas Jonathan Levin argued in the late 1950s and early 1960s that guano was fully an enclave economy, Shane Hunt tried to stress its domestic impact. However, he acknowledged that, although guano exports increased domestic demand, they did not increase domestic production and productivity. Guano brought an influx of cash in the economy, but this monetary impact did not dramatically change the structures of production. One of the reasons for this, of course, was that a great amount of the guano revenues financed the growing foreign debt, which, finally, could not be paid, precipitating the crisis of 1873–74.[19]

By 1887, after the guano boom and the War of the Pacific (itself in part a political and military consequence of the guano crisis and the rush to replace guano with nitrate deposits located in the borderlands between Peru, Bolivia, and Chile), the Peruvian government had "to retrieve from circulation 96,600,000 soles in paper money and pay for it in gold in order *to burn them.*"[20] This shows how dramatic was the monetary, fiscal, and, in general, economic crisis that the guano debacle precipitated. The Peruvian foreign debt by that time, another result of the guano boom and bust, was estimated at £51 million. This forced the Peruvian government in 1889 to sign the Grace Contract to appease its foreign creditors.[21] Thus the capital generated by the guano economy, despite its highly inflationary multiplying effect, never fully transformed the Peruvian economy and society. The capital generated by guano came easily and went easily. In the mining sector, on the other hand, capital for most of the century was insufficient.

If productive capital was scarce in the nineteenth-century Peruvian mining economy, what were the conditions for technological innovations? The most significant technological innovations were, of course, the use of steam engines for draining mines and, more important still, the railroad. Although imported steam engines, brought to Cerro de Pasco at great expense, were used to drain the mines, the results were disappointing. The socavones (draining adits built mostly by forced Indian peasant labor) were more efficient and productive.[22] Future research might profitably explore the feasibility in the nineteenth century of other imported technologies, fuels, and power sources, such as electric power and oil in the late nineteenth and early twentieth centuries; however, these technological innovations have to do largely with the transition to, and predominance of, copper mining.

In chapters 4 and 5 I dealt with the impact of the railroad on the national economy. Railroads were the other most significant technological innovation of the nineteenth century. The conclusion drawn from these chapters in terms of its effect on mining production is mixed. Railroads were an extremely costly new technology imported from abroad which did not have an outstanding impact on productivity. The railroad in Peru instead disrupted the intimate relationship between the mining sector and its transportation dimension. It certainly created unemployment. Muleteering was severely curtailed by the railroad in a dialectic of coexistence and conflict. The building of highways and the introduction of trucks in the twentieth century consigned even more arrieros to the margins of trade and transportation, where they began to resemble relics from the past. The 1870s, the most intense period of railroad building in nineteenth-century Peru, was the time of the first frontal attack on the enduring system of arriería. However, muleteering is still alive and well in Peru and is still tied, as it was more strongly in the nineteenth century, to the peasant social economy. Railroads, then, displaced one system of transportation and trade for another, which was faster and, perhaps, more efficient. But railroads did not create a technological revolution in silver mining, nor an immediate increase in output. They did, however, lay the groundwork for the penetration of foreign capital, the consolidation of larger mining companies, and the increase in the scale and the change in the logic of mining production with the predominance of copper.

Institutional and cultural barriers contributed to the slow pace of domestic capital formation and the lack of a direct and positive impact of technological innovations. Peru, as the rest of Iberian America, inherited many severe constraints from Spanish and Portuguese colonialism for the development of capitalism, among them many institutional settings that did not promote domestic, individualistic, self-reliant business practices and ethics.[23] On the contrary, as many authors have previously mentioned, corporatist, legalistic, conservative, church-oriented restrictions remained.[24] We have seen that the legal framework for mining in Peru remained the same as the 1786 Ordenanzas de Minería, established by the Spanish king and his

Council of the Indies during the Bourbon reforms. Mining deputies, the Mining Tribunal Court, and many other corporatist and guild-based institutions, or *fueros,* the remnants of Spanish times, were preserved in the nineteenth century, at least until the 1870s, and most of them—it could be argued—were still in place much later, until the new Mining Code of 1900 was established.

If business conditions did not promote an upsurge in mining, the world of labor was even more restricted. As this book clearly shows, the world of labor was the world of the Andean peasantry, a large social class and ethnic nation of small landholders living mostly in well-organized rural communities with legal rights established from colonial times. Mine workers were largely temporary laborers who went to the mines to obtain cash. Many were agriculturalists who kept their land and went back constantly to their communities. These were not specialized workers with years of mining experience but rather unskilled porters and diggers (barreteros). The mining industry in nineteenth-century Peru did not train or educate workers. And it did not create a modern mining proletariat. On the contrary, it reinforced a status quo of continuous exploitation of peasants' labor, using old forms of labor relations that certainly did not imply their proletarianization.

Why were mineowners or mining officials in the government not interested in training mine workers, and hence promoting mining development? Mineowners and the mining industry were largely motivated by the extraction of rents (from labor or from the abundance and quality of virgin natural resources), rather than the modern creation and pursuit of profit. Modern economic concepts, such as productivity or efficiency, were in general out of the minds, or at least out of reach, of government officials, mineowners, and, particularly, merchants. As I have shown, there were some initiatives in this direction, such as Juan Francisco Izcue and the Pflucker family in Morococha in the 1840s trying to produce copper, or smelting plants being established after the War of the Pacific. But the trend was to ignore economic ideas such as productivity gains or efficiency. The mining rent came from the use of cheap and forced peasant labor, but also from the abundance and wealth of a natural resource: ore deposits, especially of silver. This topic takes us into the third dimension of analysis of the mining sector in nineteenth-century Peru: the availability, abundance, and high quality of its natural resources. The observations for Peru are to some extent valid for the rest of Latin America, a continent full of natural resources largely unused during the past century. When these natural resources were actually exploited (minerals, oil, even some agricultural crops), they were largely extracted and exported as raw materials without any processing or value-added design.

Minerals were abundant and relatively easy to obtain in nineteenth-century Peru. Traditional technologies, inherited from colonial times, persisted until the end of the century. There was then, as a recent book has put it, "a subsidy from nature."[25] Economic costs, or the costs of organizing production, were relatively low; therefore profits— rather rents—tended to be high. Most of these profits, or the rent obtained from mining production, were simply the result of the extraction of a natural good that suddenly acquired a high commercial value because of international or national market conditions. In the Peruvian case this "free" natural resource was silver for most of the century, afterward the abundance and quality of copper deposits. Peru, like much of Latin America, was (and is) full of natural resources of limited domestic commercial value. The expansion of world capitalism and the commodity form in the nineteenth century, after the European Industrial Revolution and the frantic search for cheap raw materials, turned these abundant natural resources into highly prized commodities. The aphorism of the Italian traveler Antonio Raimondi was apt particularly for its mineral metaphor: "Peru is a beggar sitting on a bench of gold" *(El Perú es un mendigo sentado en un banco de oro)*. According to Raimondi, Peru was definitively a mining country with a lot of unfulfilled potential.[26]

This observation could be extended to Latin America as a whole. The abundance of valuable natural resources was mostly a subsidy of nature to the domestic and international economic process. The economic history of Latin America in the nineteenth and twentieth centuries is based on the easy availability of high-quality natural resources that could be exploited commercially at low cost. Therefore profits, coming from the use of these natural resources put in the market, tended to be high. This historical phenomenon has been called by some economists in the twentieth century "export economies of raw materials," economies organized to supply foreign markets for cheap raw materials without these materials being articulated to the domestic development of the supplier nations. The transition from silver mining to copper in late-nineteenth-century Peru produced changes in this kind of economy and increased the openness of the export sector. Peru went from small exports of precious metals that had, in their processing, first circulated domestically, to large and direct exports of a cheap and abundant raw material, copper.

Allow me then to end this work with a simple economic equation. The Mining Sector Product (MSP) was the result of the combination of Capital (C), Labor (L), and Natural Resources (NR). Thus,

$$MSP = C + L + NR$$

As I have said, the capital formation rate in the mining sector was a function (f) of the use of technological innovations (ti) not always appropriated, and institutional, cultural, and legal constraints (ilc). Furthermore, capital was rather scarce and expensive, whereas labor and natural resources were abundant and inexpensive. The equation thus becomes:

$$MSP = \quad C\,f_{ti}\,f_{ilc} \quad + \quad L \quad\quad + \quad NR$$

scarce	abundant	abundant
expensive	inexpensive	inexpensive

How can development be created in an economy in which there is an abundance of cheap natural resources and labor, but the catalyst that increases productivity is scarce and expensive? There was no interest in promoting skilled labor nor in modernizing technologically. The incentives, on the contrary, tended to maintain the social and economic status quo, while natural resources (mineral deposits or, in other cases, land) were still readily available at low cost. Such an economy tends to be static rather than dynamic.

This study of the evolution of the mining sector in nineteenth-century Peru is an example that could be extended to other cases in Latin America and in the world economy. Several countries in Latin America have also been paises mineros, mining countries—Bolivia, Chile, and Mexico, for example. Like Peru, these countries have had a long history of mining that started in pre-Columbian times. In the first half of the nineteenth century, as we have seen in chapter 2, Mexico was well ahead of Peru in silver production, whereas Bolivia and Chile lagged behind. This situation changed in the second half of the century, when silver production in Bolivia and Chile outstripped that of Peru. But the important feature here is that in all these countries mining became an open export economy toward the end of the century.

I have stressed in this book the opposite directions that mining could take for the development of a country in modern times, focusing specifically on Peru in the nineteenth century. Mining could serve either as an export economy or to support domestic development, particularly expanding the internal market and principally the purchasing power of the lower classes, which for nineteenth-century Peru also meant the incorporation of Andean peasants as full-fledged citizens and as key economic actors in the nation. Mining could also take both directions, in which case tensions develop when one pulls harder than the other. Silver mining tended to foster domestic markets in nineteenth-century Peru, whereas copper encouraged more of an open export economy. Historically, export economies have tended to prevail in Latin America. They have served the international economies, particularly Europe and the United States, rather than their very own, and they took a hold in every Latin American country sometime during the nineteenth century.

We have heard some voices from nineteenth-century Peru. A historian has organized them into a text, this *opera aperta* that now closes its pages. The historian writes in the present about the past in order to understand further our present condition and, therefore, to foresee the future with increased clarity. It is now up to you, dear reader, to make up your mind.

Abbreviations

AAEP	Archive des Affaires Etrangères de Paris.
ACNL	Archivo del Congreso Nacional, Lima.
ADRMCP	Archivo de la Dirección Regional de Minería del Cerro de Pasco, Cerro de Pasco, Perú.
ADRMH	Archivo de la Dirección Regional de Minería de Huancayo, Huancayo, Perú.
AFA	Archivo del Fuero Agrario, Lima.
AGN	Archivo General de la Nación, Lima.
ALCMP	Archivo Legal de la Empresa Minera Centro-Min Perú, Lima.
ALG	Sección Algolán in AFA.
AMNH	Archivo del Museo Nacional de Historia, Lima.
AMRE	Archivo del Ministerio de Relaciones Exteriores, Lima.
AUNI	Archivo de la Universidad Nacional de Ingeniería, Lima.
BN	Biblioteca Nacional, Lima.
CCC	Correspondance Commerciale et Consulaire.
FO	Foreign Office.
PRO	Public Record Office, London.
SCM	Sección Casa de Moneda, in AGN.
SHMH	Sección Histórica del Ministerio de Hacienda, in AGN.

Notes

Chapter 1

1. See, for example, Gilbert M. Joseph and Daniel Nugent, eds., *Everyday Forms of State Formation: Revolution and the Negotiation of Rule in Modern Mexico* (Durham, N.C.: Duke University Press, 1994); Florencia E. Mallon, *Peasant and Nation: The Making of Postcolonial Mexico and Peru* (Berkeley: University of California Press, 1995); and particularly, for a more extreme case, Marjorie Becker, *Setting the Virgin on Fire: Lázaro Cárdenas, Michoacán Peasants, and the Redemption of the Mexican Revolution* (Berkeley: University of California Press, 1995).

2. For an assessment of the debate in the historical discipline in the United States see Dorothy Ross, "Grand Narrative in American Historical Writing: From Romance to Uncertainty," *American Historical Review* 100, 3 (June 1995): 651–77; Gordon S. Wood, "A Century of Writing Early American History: Then and Now Compared; or, How Henry Adams Got It Wrong." *American Historical Review* 100, 3 (June 1995): 678–96; and Dominick LaCapra, "History, Language, and Reading: Waiting for Crillon," *American Historical Review* 100, 3 (June 1995): 799–828. For a Peruvian assessment see *Revista andina* 9, 1 (Centro Bartolomé de las Casas, Cusco, 1991), particularly 123–259, which contains discussions on Andean history and etnohistory, early and late colonial history, and nineteenth-century historiography by Henrique Urbano, Gabriela Ramos, Pedro Guibovich, Susana Aldana Rivera, and Nelson Manrique.

3. Adam Smith, for example, published his *Theory of Moral Sentiments* (1759) before writing *An Inquiry into the Nature and Causes of the Wealth of Nations* (1776).

4. Aristotle, *Metaphysics,* trans. Richard Hope (Ann Arbor: University of Michigan Press, 1960), 3.

5. Antonio Gramsci, *El materialismo histórico y la filosofía de Benedetto Croce* (Buenos Aires: Nueva Visión, 1972), 7. See also Gramsci, *Selections from the Prison Notebooks,* ed. and trans. Q. Hoare and G. Nowell Smith

(New York: International Publishers, 1971), 347 and 419. Parallels have been established between Gramsci and Peruvian socialist intellectual and activist José Carlos Mariátegui. For example in Robert Paris, "El marxismo de Mariátegui," *Aportes* 17 (Revista de Estudios Latinoamericanos, Paris, July 1970); and Alberto Flores Galindo, *La agonía de Mariátegui* (Lima: DESCO, 1980).

6. I refer to Garcilaso de la Vega because he was the most important Peruvian writer, from Inca and mestizo origins, in the sixteenth century, whereas Mariano Eduardo de Rivero y Ustáriz filled that role, in mining, scientific, and archeological topics, for the nineteenth century. Garcilaso de la Vega's most important work is *Comentarios reales de los Incas*, 3 vols. (Lima: Ediciones Peisa, 1973); this work includes an introduction and commentary by José Durand. Gold, silver, and mercury mines and their work are treated in vol. 3, 134-39. On Mariano de Rivero, see chapters 2 and 3.

7. See Peter Novick, *That Noble Dream: The "Objectivity Question" and the American Historical Profession* (Cambridge: Cambridge University Press, 1988).

8. See for example Chester G. Starr, *A History of the Ancient World*, 4th ed. (New York: Oxford University Press, 1991, fourth edition), 75-77, 94-96, 124-29. A particular case can be seen in Fundación Rio Tinto, *La comarca de Rio Tinto: Un territorio de mina* (Huelva, Spain: Fundación Rio Tinto, 1994), particularly p. 37.

9. Earl J. Hamilton, *The American Treasure and the Price Revolution in Spain, 1501-1660* (Cambridge: Harvard University Press, 1934). For a recent revisionist balance of the Latin American contribution in gold and silver remittances to the development of the European economies, see Michel Morineau, *Incroyables gazettes et fabuleux métaux: Les rétours des trésors américaines d'après les gazettes hollandaises (seizième-dix-huitième siècles)* (Paris and London: Editions de la Maison des Sciences de l'Homme and Cambridge University Press, 1985).

10. See T. S. Ashton, *The Industrial Revolution, 1760-1830* (Oxford: Oxford University Press, 1948); also Paul Mantoux, *La révolution industrielle aux dix-huitième siècle* (Paris: Editions Génin, 1959); David S. Landes, *The Unbound Prometheus: Technological Change and Industrial Development in Western Europe from 1750 to the Present* (Cambridge: Cambridge University Press, 1969), particularly pp. 41-192; and, more recently, Peter N. Stearns, *Interpreting the Industrial Revolution* (Washington, D.C.: American Historical Association, 1991).

11. See, for example, J. Gallagher and R. Robinson, "The Imperialism of Free Trade," *Economic History Review*, 2d series, 6 (1953): 1-15; D. C. M. Platt, "The Imperialism of Free Trade: Some Reservations," *Economic History Review* 21 (1968): 296-306. See also D. C. M. Platt, *Latin America and British Trade, 1806-1914* (New York: Harper and Row, 1973).

12. D. C. M. Platt, "Dependency in Nineteenth-Century Latin America: An Historian Objects," *Latin American Research Review* 15, 1 (1980): 113-30; Barbara Stein and Stanley Stein, "D. C. M. Platt, the Anatomy of 'Autonomy,'" *Latin American Research Review* 15, 1 (1980): 131-46.

13. See Heraclio Bonilla, *Guano y burguesía en el Perú* (Lima: Instituto de Estudios Peruanos, 1974); Paul Gootenberg, "The Social Origins of Protectionism and Free Trade in Nineteenth-Century Lima," *Journal of Latin American Studies* 14, 2 (1982): 329-58. The reply of Bonilla can be found in *Guano y burguesía*, 7-12. A study that also de-emphasizes the role, intervention, and earnings of a British firm in the nineteenth-century Peruvian guano economy, and therefore supports the "autonomy" view of Latin America, is William M. Mathew, "Anthony Gibbs and Sons, the Guano Trade and the Peruvian Government, 1842-1861," in *Business Imperialism, 1840-1930: An Inquiry Based on British Experience in Latin America,* ed. D. C. M. Platt (Oxford: Clarendon Press, 1977), 337-70. See also William M. Mathew, *The House of Gibbs and the Peruvian Guano Monopoly* (London: Royal Historical Society, 1981).

14. See Paul Gootenberg, *Between Silver and Guano: Commercial Policy and the State in Postindependence Peru* (Princeton: Princeton University Press, 1989), particularly 18-25; Florencia E. Mallon, *The Defense of Community in Peru's Central Highlands: Peasant Struggle and Capitalist Transition, 1860-1940* (Princeton: Princeton University Press, 1983), particularly 168-234. See also Paul Gootenberg, *Tejidos y harinas, corazones y mentes: El imperialismo norteamericano del libre comercio en el Perú, 1825-1840* (Lima: Instituto de Estudios Peruanos, 1989).

15. The same is true for some of Gootenberg's works, particularly *Tejidos y harinas.* For a different view on the impact of free and foreign trade on a Peruvian regional economy, see Heraclio Bonilla, Lía del Río, and Pilar Ortiz de Zevallos, "Comercio libre y crisis de la economía andina: El caso del Cuzco," *Histórica* 2, 1 (Pontificia Universidad Católica del Perú, Lima, July 1978): 1-25.

16. On the question of linkages and the interdependence that an "underdeveloped" economy has to have with other economic sectors, see Albert

O. Hirschman, *The Strategy of Economic Development* (New Haven: Yale University Press, 1958); although he sees these backward and forward linkages in terms of the industrial sector and the industrialization process. See particularly 98–119.

17. As examples of a field in current expansion, see Dolores Avila, Inés Herrera, and Rina Ortiz, ed., *Empresarios y política minera* (Mexico City: Instituto Nacional de Antropología e Historia, 1992); Inés Herrera Canales and Rina Ortiz Peralta, eds., *Minería americana colonial y del siglo diecinueve* (Mexico City: Instituto Nacional de Antropología e Historia, 1994); and Inés Herrera Canales, Rina Ortiz Peralta, María Eugenia Romero Sotelo, and José Alfredo Uribe Salas, *Ensayos sobre minería mexicana, siglos dieciocho al veinte* (Mexico City: Instituto Nacional de Antropología e Historia, 1996).

18. See Pierre Vayssiere, *Un siecle de capitalisme minier au Chili, 1830–1930* (Toulouse: Centre National de la Recherche Scientifique, 1980); Steven S. Volk, "Crecimiento sin desarrollo: Los propietarios mineros chilenos y la caida de la minería en el siglo diecinueve," in Herrera Canales and Ortiz Peralta, eds., *Minería americana colonial*, 69–118; and Marcela Orellana Muermann and Juan G. Muñoz Correa, eds., *Mundo minero: Chile, siglos diecinueve y veinte* (Santiago: Universidad de Santiago de Chile, 1991).

19. See Antonio Mitre, *Los patriarcas de la plata: Estructura socioeconómica de la minería boliviana en el siglo diecinueve* (Lima: Instituto de Estudios Peruanos, 1981); and Herbert S. Klein, *Bolivia: The Evolution of a Multi-Ethnic Society* (New York: Oxford University Press, 1982), particularly 149–87.

20. Gootenberg, *Between Silver and Guano.*

21. Paul Gootenberg, "*Carneros y Chuño:* Price Levels in Nineteenth-Century Peru," *Hispanic American Historical Review* 70, 1 (1990): 1–56; "Population and Ethnicity in Early Republican Peru: Some Revisions," *Latin American Research Review* 26, 3 (1991): 109–57; *Imagining Development: Economic Ideas in Peru's "Fictitious Prosperity" of Guano, 1840–1880* (Berkeley: University of California Press, 1993).

22. Christine Hünefeldt, *Paying the Price of Freedom: Family and Labor among Lima's Slaves, 1800–1854* (Berkeley: University of California Press, 1994); see also Hünefeldt, *Los Manuelos: Vida cotidiana de una familia negra en la Lima del siglo diecinueve: Una reflexión histórica sobre la esclavitud urbana* (Lima: Instituto de Estudios Peruanos, 1992).

23. Peter Blanchard, *Slavery and Abolition in Early Republican Peru* (Wilmington, Del.: Scholarly Resources, 1992). See also the review of this work by Christine Hünefeldt in *Hispanic American Historical Review* 73, 4 (1993): 711–13.

24. Frederick P. Bowser, *The African Slave in Colonial Peru, 1524–1650* (Stanford: Stanford University Press, 1974).

25. See, for example, Robert G. Keith, *Conquest and Agrarian Change. The Emergence of the Hacienda System on the Peruvian Coast* (Cambridge: Harvard University Press, 1976); Manuel Burga, *De la encomienda a la hacienda capitalista: El valle del Jequetepeque del siglo dieciseis al veinte* (Lima: Instituto de Estudios Peruanos, 1976); and Susan E. Ramírez, *Provincial Patriarchs: Land Tenure and the Economics of Power in Colonial Peru* (Albuquerque: University of New Mexico Press, 1986).

26. See, for example, Peter F. Klarén, *Modernization, Dislocation, and Aprismo: Origins of the Peruvian Aprista Party, 1870–1932* (Austin: University of Texas Press, 1973), particularly 3–64; Pablo Macera, "Las plantaciones azucareras andinas (1821–1875)" (original edition, 1974), in *Trabajos de historia* (Lima: Instituto Nacional de Cultura, 1977), vol. 4, 9–307; Burga, *Encomienda a la hacienda capitalista,* particularly 165–223; finally, more recently, Michael J. Gonzales, *Plantation Agriculture and Social Control in Northern Peru, 1875–1933* (Austin: University of Texas Press, 1985).

27. Carlos Aguirre, *Agentes de su propia libertad: Los esclavos y la desintegración de la esclavitud, 1821–1854* (Lima: Pontificia Universidad Católica del Perú, 1993).

28. Alfonso W. Quiroz, *Domestic and Foreign Finance in Modern Peru, 1850–1950: Financing Visions of Development* (Pittsburgh: University of Pittsburgh Press, 1993). See also Carlos Camprubí, *Historia de los bancos en el Perú (1860–1879)* (Lima: Editorial Lumen, 1957); and *El banco de la emancipación* (Lima: Imprenta P. L. Villanueva, 1960).

29. Alfonso W. Quiroz, *Banqueros en conflicto: Estructura financiera y economía peruana, 1884–1930* (Lima: Universidad del Pacífico, 1990).

30. Nils Jacobsen, *Mirages of Transition: The Peruvian Altiplano, 1780–1930* (Berkeley: University of California Press, 1993).

31. Mallon, *Defense of Community.*

32. Ernesto Yepes del Castillo, *Perú 1820–1920: Un siglo de desarrollo capitalista* (Lima: Instituto de Estudios Peruanos and Campodónico Ediciones, 1972); Julio Cotler, *Clases, estado, y nación en el Perú* (Lima: Instituto de Estudios Peruanos, 1978).

33. Pablo Macera, "La historia en el Perú: Ciencia e ideología," *Amaru* 6 (Revista de Artes y Ciencias, Universidad Nacional de Ingeniería, Lima, April–June 1968). Also reproduced in Macera, *Trabajos de historia*, 4 vols. (Lima: Instituto Nacional de Cultura, 1977), 1:3–20. See also Macera, "Explicaciones," in *Trabajos de Historia* 1:vii–lxxvi; Heraclio Bonilla, "The New Profile of Peruvian History," *Latin American Research Review* 16, 3 (1981): 210–24; José Deustua, "Sobre movimientos campesinos e historia regional en el Perú moderno: Un comentario bibliográfico," *Revista andina* 1, 1 (Cusco, 1983): 219–40; Fred Bronner, "Peruvian Historians Today: Historical Setting," *Americas* 43, 3 (1986): 245–77; Christine Hünefeldt, "Viejos y nuevos temas de la historia económica del siglo diecinueve," in *Las crisis económicas en la historia del Perú*, ed. Heraclio Bonilla (Lima: Centro Latinoamericano de Historia Económica y Social and Fundación Friedrich Ebert, 1986), 33–60; and Nelson Manrique, "La historiografía peruana sobre el siglo diecinueve," *Revista andina* 9, 1 (1991): 241–59.

34. Some of the main works of Jorge Basadre are: *La Multitud, la ciudad, y el campo en la historia del Perú* (Lima: Ediciones Treintaitrés and Mosca Azul Editores, 1980; original edition, 1929); *Perú: Problema y posibilidad* (Lima: Banco Internacional del Perú, 1978; original edition, 1931); *Historia de la república del Perú, 1822–1933*, 6th ed., 17 vols. (Lima: Editorial Universitaria, 1968–70); and *Introducción a las bases documentales para la historia de la república del Perú con algunas reflexiones* 2 vols. (Lima: Ediciones P. L. Villanueva, 1971).

35. Many of Pablo Macera's works are compiled in his *Trabajos de Historia*. See also Macera, "Mapas coloniales de haciendas cuzqueñas," Seminario de Historia Rural Andina, Universidad Nacional Mayor de San Marcos, Lima, 1968 (mimeo); and more recently, on price history, Macera et al., *Los precios del Perú, siglos dieciseis-diecinueve: Fuentes*, 3 vols. (Lima: Banco Central de Reserva del Perú, 1992). Some of his works on colonial sexuality, language, and, in general, cultural studies are: "Sexo y coloniaje," in *Trabajos de historia*, 3:297–352; "Lenguaje y modernismo peruano del siglo dieciocho," in *Trabajos de historia*, 2:9–77; and "El probabilismo en el Perú del siglo dieciocho," in *Trabajos de historia*, 2:79–137. See more references to Basadre's and Macera's works in the following chapters.

36. Pablo Macera, "El arte mural cuzqueño, siglos dieciseis–veinte," *Apuntes* 2, 4 (Revista Semestral de Ciencias Sociales, Universidad del Pacífico, Lima, 1975); Macera, *Pintores populares andinos* (Lima: Fondo del Libro del Banco de los Andes, 1979); Macera, *Retablos andinos* (Lima: Uni-

NOTES TO PAGES 10-11

versidad Nacional Mayor de San Marcos, 1981); Macera, "Arte y lucha social: Los murales de Ambaná (Bolivia)," *Allpanchis* 15, 17/18 (Instituto de Pastoral Andina, Cusco, 1981): 23–40.

37. Compare, for example, the treatment of mining in the sixteenth, seventeenth, and nineteenth centuries in Carlos Sempat Assadourian, Heraclio Bonilla, Antonio Mitre, and Tristan Platt, *Minería y espacio económico en los Andes, siglos dieciseis-veinte* (Lima: Instituto de Estudios Peruanos, 1980), particularly 20–44 and 46–51. On the Potosí mining boom of the sixteenth and seventeenth centuries, see Peter Bakewell, *Miners of the Red Mountain: Indian Labor in Potosí, 1545–1650.* (Albuquerque: University of New Mexico Press, 1984); and Jeffrey A. Cole, *The Potosí Mita, 1573–1700: Compulsory Indian Labor in the Andes* (Stanford: Stanford University Press, 1985).

38. See, for example, Assadourian et al., *Minería y espacio económico,* 46.

39. John R. Fisher, *Minas y mineros en el Perú colonial, 1776–1824* (Lima: Instituto de Estudios Peruanos, 1977), Spanish translation of *Silver Mines and Silver Miners in Colonial Peru, 1776–1824* (Liverpool: University of Liverpool, 1977); Enrique Tandeter, *Coercion and Market: Silver Mining in Colonial Potosí, 1692–1826* (Albuquerque: University of New Mexico Press, 1993).

40. See, for example, Yepes del Castillo, *Perú,* 137–40; Cotler, *Clases, estado, y nación,* 125–6; Mallon, *Defense of Community,* 129; Lawrence A. Clayton, *Grace: W. R. Grace and Co., The Formative Years, 1850–1930* (Ottawa, Ill.: Jameson Books, 1985), particularly chap. 7, 141–75; and more specifically, Rory Miller, "The Making of the Grace Contract: British Bondholders and the Peruvian Government, 1885–1890," *Journal of Latin American Studies* 8 (1976): 73–100.

41. According to economist Shane Hunt, the Peruvian national income or national product for one fiscal year, 1876–1877, represented some 235,518,000 silver soles in current prices. This meant a per capita income of 87 soles per year. See Shane Hunt, "Growth and Guano in Nineteenth-Century Peru" (discussion paper no. 34, Woodrow Wilson School, Princeton, 1973), table 14 and app. Carlos Boloña has elaborated other estimations for the end of the nineteenth and the first half of the twentieth centuries. He estimates a GNP of 196 million current soles for the year 1900. See Carlos Boloña, "Tariff Policies in Peru, 1880–1980" (Ph.D. dissertation, Oxford University, 1981); also Boloña, "Perú: Estimaciones preliminares del producto nacional, 1900–1942," *Apuntes* 13 (Revista de Ciencias Sociales, Universidad del Pacífico, Lima, 1983): 3–13. For different estimates of and

data on exports, wages, and employment in nineteenth-century Peru, see Macera, "Plantaciones azucareras," 99–100, table 6, and pp. 150–228; for estimates of the Lima economy between 1830 and 1861 based on *"matrícula de patentes,"* see Gootenberg, *Between Silver and Guano,* app. 2, table 2.1 See also Paul Gootenberg, "Merchants, Foreigners, and the State: The Origins of Trade Policies in Post-Independence Peru" (Ph.D. dissertation, University of Chicago, 1985); finally, see Quiroz, *Domestic and Foreign Finance,* app. A, table A.1, 219–21, for estimates of financial assets compared to GNP for certain years between 1850 and 1965.

42. I have to insist here on the importance of agricultural activities and the rural population, whether living in haciendas or peasant communities, for nineteenth-century Peru. Hunt, in *Growth and Guano,* estimates that for 1876–1877 86 percent of the labor force was engaged in rural occupations. I have estimated elsewhere, following the 1876 national census (also an important source for Hunt's and Macera's works), that 83 percent of the total Peruvian population of 2,670,000 lived in the countryside. See José Deustua, "Mining Markets, Peasants, and Power in Nineteenth-Century Peru," *Latin American Research Review,* 29, 1 (1994): 29–54, particularly 42, and the sources mentioned in note 52. Finally, I have also made preliminary calculations of the value of the domestic agricultural production of corn and potatoes for late-nineteenth-century Peru, finding it as important as that of silver and guano. See José Deustua, "Producción minera y circulación monetaria en una economía andina: El Perú del siglo diecinueve," *Revista andina* 4, 2 (Cusco, 1986): 319–54, particularly 341–42.

43. See Fernand Braudel, *Ecrits sur l'histoire.* (Paris: Flammarion, 1969), particularly part 2, "L'histoire et les autres sciences de l'homme" (History and the other human sciences), 39–235. The English edition, with some changes, is Fernand Braudel, *On History,* trans. Sarah Matthews (Chicago: University of Chicago Press, 1980).

44. Warren J. Samuels, ed., *The Economy as a System of Power* (New Brunswick, N.J.: Transaction Books, 1979).

45. Philip A. Klein, "Economics: Allocation or Valuation?" in Samuels, ed., *Economy as a System,* 7–33.

46. Eric J. Hobsbawm, *The Age of Capital, 1848–1875* (London: Abacus, 1977), 44.

47. For an interesting discussion, although uneven, of the role of the state and market in a Latin American historical context, see Joseph L. Love and Nils Jacobsen, eds., *Guiding the Invisible Hand. Economic Liberalism and the State in Latin American History* (New York: Praeger, 1988).

48. See William M. Reddy, *The Rise of Market Culture: The Textile Trade and French Society, 1750-1900* (Cambridge and Paris: Cambridge University Press and Editions de la Maison des Sciences de l'Homme, 1984).

49. See John V. Murra, *La organización económica del estado inca* (Mexico City: Siglo Veintiuno Editores, 1978); Murra, *Formaciones económicas y políticas del mundo andino* (Lima: Instituto de Estudios Peruanos, 1975); María Rostworowski, *Etnía y sociedad: Costa peruana prehispánica* (Lima: Instituto de Estudios Peruanos, 1977); Giorgio Alberti and Enrique Mayer, eds., *Reciprocidad e intercambio en los Andes peruanos* (Lima: Instituto de Estudios Peruanos, 1974); Mallon, *Defense of Community*, particularly 24-41; and Nathan Wachtel, *Le rétour des ancêtres: Les Indiens urus de Bolivie, vingtième-seizième siecles: Essai d'histoire régressive* (Paris: Gallimard, 1990).

50. Steve J. Stern, *Peru's Indian Peoples and the Challenge of Spanish Conquest. Huamanga to 1640*, 2d ed. (Madison: University of Wisconsin Press, 1993); Olivia Harris, Brooke Larson, and Enrique Tandeter, eds., *La participación indígena en los mercados surandinos: Estrategias y reproducción social, siglos dieciseis al veinte.* (La Paz: Centro de Estudios de la Realidad Social y Económica, 1988); for an early discussion of the contrast between Murra's, Rostworowski's, and Valcárcel's views on Andean ethnohistory, José Deustua, "Derroteros de la etnohistoria en el Perú," *Allpanchis* 14, 15 (Instituto de Pastoral Andina, Cusco, 1980): 173-78.

51. See Eric R. Wolf, *Peasants* (Englewood Cliffs, N.J.: Prentice Hall, 1966); and Theodor Shanin, ed., *Peasants and Peasant Societies* (Baltimore: Penguin, 1971), for the importance of peasant societies without or with a small presence of the market. Of course, see also Alexander V. Chayanov, *The Theory of Peasant Economy* (Madison: University of Wisconsin Press, 1986; original Russian edition, 1923); and Daniel Thorner, "L'économie paysanne: Concept pour l'histoire économique," *Annales: Economies, sociétés, civilizations* 19, 3 (Paris, May–June, 1964): 417-32.

52. I am referring, of course, to Vladimir Ilich Lenin, *The Development of Capitalism in Russia* (Moscow: Progress Publishers, 1974; original edition, 1899). See also Deustua, "Mining Markets, Peasants," particularly 29-33.

53. Eric R. Wolf, "Types of Latin American Peasantry: A Preliminary Analysis," *American Anthropologist* 57 (June 1955): 452-71; and Wolf: "Closed Corporate Peasant Communities in Mesoamerica and Central Java," *Southwestern Journal of Anthropology* 13 (1957): 1-18.

54. In Clodoaldo Alberto Espinosa Bravo, *El hombre de Junín frente a su paisaje i a su folklore*, 2 vols. (Lima: Talleres Gráficos P. L. Villanueva,

1967): 1:47. On the historical, social, and cultural background of Indian and mestizo songs in the central sierra of Peru, see José María Arguedas, "Simbolismo y poesía de dos canciones populares quechuas" and "La canción popular mestiza e india en el Perú," as well as other contributions in *Indios, mestizos, y señores* (Lima: Editorial Horizonte, 1985); see also Arguedas, "Evolución de las comunidades indígenas: El valle del Mantaro y la ciudad de Huancayo: Un caso de fusión de culturas no comprometidas por la acción de las instituciones de origen colonial," *Revista del Museo Nacional* 26 (Lima, 1957). Finally, for a preliminary analysis of the relationships between "fiestas" and class and ethnic conflicts, see Nelson Manrique, *Yawar Mayu: Sociedades terratenientes serranas, 1879–1910* (Lima: DESCO and Instituto Francés de Estudios Andinos, 1988), particularly 21–50.

55. See Franklin Pease, *Del Tawantinsuyu a la historia del Perú* (Lima: Instituto de Estudios Peruanos, 1978); and Alberto Flores Galindo, *Buscando un Inca: Identidad y utopía en los Andes* (Havana: Casa de las Américas, 1986), for continuities of the Andean tradition, identity, and, according to Flores Galindo, utopia. For elaboration on the resistance-adaptation dialectic, see Steve Stern, ed., *Resistance, Rebellion, and Consciousness in the Andean Peasant World, Eighteenth to Twentieth Centuries* (Madison: University of Wisconsin Press, 1987); and particularly in it Stern, "New Approaches to the Study of Peasant Rebellion and Consciousness: Implications of the Andean Experience," 3–25.

56. Nathan Wachtel, *La vision des vaincus: Les Indiens du Pérou devant la conquête espagnole, 1530–1570* (Paris: Gallimard, 1971); see also Ruggiero Romano, *Les mécanismes de la Conquête coloniale: Les conquistadores* (Paris: Flammarion, 1972).

57. Carlos Sempat Assadourian, *El sistema de la economía colonial: Mercado interno, regiones y espacio económico* (Lima: Instituto de Estudios Peruanos, 1982); see also Assadourian, "La producción de la mercancía dinero en la formación del mercado interno colonial: El caso del espacio peruano, siglo dieciseis," in *Ensayos sobre el desarrollo económico de México y América Latina (1500–1975)*, ed. Enrique Florescano (Mexico City: Fondo de Cultura Económica, 1979), 223–92; finally, see Luis Miguel Glave, *Trajinantes: Caminos indígenas en la sociedad colonial, siglos dieciseis–diecisiete* (Lima: Instituto de Apoyo Agrario, 1989).

58. See Enrique Tandeter and Nathan Wachtel, "Conjonctures inverses: Le mouvement des prix à Potosí pendant le dix-huitième siècle," *Annales: Economies, sociétés, civilizations* 38, 3 (Paris, May–June 1983): 549–613; and

Enrique Tandeter, Vilma Milletich, María Matilde Ollier, and Beatriz Ruibal, "El mercado de Potosí a fines del siglo dieciocho," in Harris et al., eds., *Participación indígena*, 379–424.

59. I follow here a long tradition of economic thinking that has to do rather more with the German debates of the nineteenth and early twentieth centuries on economic history and the relationship between *Naturalwirtschaft* and *Geldwirtschaft* (cf. Hildebrand, Marx, Weber, Dopsch, etc.) than with more market functionalist, marginalist, neoliberal authors, from Bentham and Marshall to Milton Friedman. See Ruggiero Romano, "Fundamentos del funcionamiento del sistema económico colonial," in *El sistema colonial en la América española*, ed. Heraclio Bonilla (Barcelona: Editorial Crítica, 1991), 239–80; see also Romano, "American Feudalism," *Hispanic American Historical Review* 64, 1 (1984): 121–34.

60. Jurgen Golte, *Repartos y rebeliones: Tupac Amaru y las contradicciones de la economía colonial* (Lima: Instituto de Estudios Peruanos, 1980); Manuel Burga and Wilson Reátegui, *Lanas y capital mercantil en el sur: La Casa Ricketts, 1895–1935* (Lima: Instituto de Estudios Peruanos, 1981).

61. Marc Bloch, *The Historian's Craft*, trans. Peter Putnam (New York: Vintage Books, 1953), 25.

Chapter 2

1. The first modern works on social and economic Peruvian history started to appear in the late 1950s, influenced by the French historical school and focusing mostly on rural agrarian themes. If I have to mention an institution it would be the Seminario de Historia Rural Andina of the Universidad de San Marcos in Lima. And I would have to name Pablo Macera. Until 1981 the seminario had published 86 titles on "rural, agrarian, Andean history." See the catalogue published in *Allpanchis* 15, 17/18 (Instituto de Pastoral Andina, Cusco, 1981), 269–74. See also Deustua, "Movimientos campesinos," particularly 220–23; and Hünefeldt, "Viejos y nuevos temas."

2. Carlos P. Jiménez, "Reseña histórica de la minería en el Perú," in *Síntesis de la minería peruana en el centenario de Ayacucho* (Lima: Imprenta Torres Aguirre, 1924), 45. All translations from Spanish to English are mine.

3. Denis Sulmont, "Historia del movimiento obrero minero metalúrgico (hasta 1970)," *Tarea*, no. 2 (Revista de Cultura, Lima, October 1980): 30.

4. Alberto Flores Galindo, *Los mineros de la Cerro de Pasco, 1900–1930* (Lima: Pontificia Universidad Católica del Perú, 1974), 23.

5. José Morales y Ugalde, *Manifestación del estado de la hacienda de la república del Perú en fin de Abril de 1827: Presentada al soberano congreso constituyente por el ciudadano encargado de la dirección del ministerio* (Lima: Imprenta Rep. por J. M. Concha, 1827), 15.

6. Johann Jakob von Tschudi, *Testimonio del Perú, 1838–1842* (Lima: Consejo Consultivo Suiza-Perú, 1966), 266.

7. Pedro C. Venturo, "Excursiones científicas: Viaje al asiento mineral del Cerro de Pasco," *Boletín de minas, industria, y construcciones,* 13, 7 (Lima, 10 August 1897): 51.

8. Fernand Braudel, *La dynamique du capitalisme* (Paris: Flammarion, 1985), 11. On the discussions of the role of economic history and quantification, see Pierre Chaunu, *Histoire quantitative, histoire sérielle* (Paris: Cahiers des Annales, 1978), a reflection inspired by Chaunu's twelve-volume masterpiece, *Séville et l'Atlantique (1504–1650): Structures et conjoncture de l'Atlantique espagnol et hispano-américain* (Paris: SEVPEN, 1955–1960); Pierre Vilar, *Crecimiento y desarrollo: Economía e historia: Reflexiones sobre el caso español,* 3d ed. (Barcelona: Editorial Ariel, 1976), a collection of his articles written during and after the elaboration of his masterpiece, *La Catalogne dans l'Espagne moderne: Recherches sur les fondements économiques des structures nationales,* 3 vols. (Paris: SEVPEN, 1962).

9. Heraclio Bonilla, "La coyuntura comercial del siglo diecinueve en el Perú," *Revista del Museo Nacional* 35 (Lima, 1967–68): 159–187; Pablo Macera and Onorio Pinto, "Estadísticas históricas del Perú: Sector minero 2 (Volumen y valor)" (Lima: Centro Peruano de Historia Económica, 1972) (mimeo); Pablo Macera, "Estadísticas históricas del Perú: Sector minero (Precios)" (Lima: Centro Peruano de Historia Económica, 1972) (mimeo); Shane Hunt, "Price and Quantum Estimates of Peruvian Exports, 1830–1962" (discussion paper no. 34, Woodrow Wilson School, Princeton, 1973); Fisher, *Minas y mineros;* Charles McArver, "Mining and Diplomacy: United States Interests at Cerro de Pasco, 1876–1930" (Ph.D. dissertation, University of North Carolina, 1977); Donald Tarnawiecki, "Crisis y desnacionalización de la minería peruana: El caso de Cerro de Pasco, 1880–1901" (tesis de economía, Pontificia Universidad Católica del Perú, Lima, 1978); José Deustua, *La minería peruana y la iniciación de la república, 1820–1840* (Lima: Instituto de Estudios Peruanos, 1986); Carlos Contreras, *Mineros y campesinos en los Andes: Mercado laboral y economía campesina en la sierra*

central, siglo diecinueve (Lima: Instituto de Estudios Peruanos, 1988); and José Deustua, "Mines, monnaie, et hommes dans les Andes: Une histoire économique et sociales de l'activité minière dans le Pérou du dix-neuvième siècle" (Thèse de doctorat, Ecole des Hautes Etudes en Sciences Sociales, Paris, 1989).

10. Mariano Eduardo de Rivero y Ustáriz, *Colección de memorias científicas, agrícolas, e industriales publicadas en distintas épocas,* 2 vols. (Brussels: Imprenta de H. Goemaere, 1857); Mateo Paz Soldán, *Geografía del Perú* (Paris: Librería de Fermin Didot, Hermanos, Hijos y Cia, 1862); Mariano Felipe Paz Soldán, *Diccionario geográfico-estadístico del Perú* (Lima: Imprenta del Estado, 1877); Maurice Du Chatenet, *Estado actual de la industria minera en el Cerro de Pasco* (Lima: Anales de la Escuela de Construcciones Civiles y de Minas, 1880).

11. Pedro Dávalos y Lisson: "La industria minera," in *El Perú* (Lima, n.p., 1900); Jiménez, "Reseña histórica."

12. On the social and institutional history of Pasco during colonial times, see César Pérez Arauco, "Anales del Cerro de Pasco. Referencias cronológicas de nuestra historia," *El pueblo,* no. 18 (Revista Cultural de Difusión Popular, Cerro de Pasco, November 1978); and Pérez Arauco, "Cerro de Pasco: Historia del pueblo mártir del Perú, siglos dieciseis, diecisiete, dieciocho, y diecinueve" (Cerro de Pasco: Edición de El Pueblo, 1980) (mimeo); Mariano Pacheco, Miguel Salcedo, and Toribio Yantas, "Pasco en la colonia, siglos dieciseis, diecisiete, y dieciocho," in *Pasco colonial* (Universidad Nacional Daniel Alcides Carrión, Cerro de Pasco, 1980), 1–38 (mimeo); Alejandro Reyes Flores, "Estudios socio-económicos de los pueblos de Pasco, siglo dieciocho," in *Pasco colonial* (Universidad Nacional Daniel Alcides Carrión, Cerro de Pasco, 1980): 39–88 (mimeo); César Espinoza Claudio and José Boza Monteverde, "Alcabalas y protesta popular: Cerro de Pasco 1780" (Lima: Universidad Nacional Mayor de San Marcos, 1981) (mimeo). See also Fisher, *Minas y mineros.* Finally there are some other references in Alejandro Reyes Flores, *Contradicciones en el Perú colonial, región central 1650–1810* (Lima: Universidad Nacional Mayor de San Marcos, 1983).

13. See Arturo Alcalde Mongrut, *Mariano de Rivero-Federico Villareal* (Lima: Editorial Universitaria, 1966); Manuel de Mendiburu, *Diccionario histórico-biográfico del Perú* (Lima: Librería e Imprenta Gil, 1934), 9:429–30; and Alberto Tauro del Pino, *Diccionario enciclopédico del Perú,* 3 vols. plus appendix (Lima: Editorial Juan Mejía Baca, 1967), 3:63.

14. Rivero y Ustáriz, *Colección de memorias,* a collection of most of his works. Some of his manuscripts (letters, reports) can be found in AGN, Sección Histórica del Ministerio de Hacienda (hereafter SHMH), PL 6, nos. 164 and 177 (1826); and PL 7, no. 225 (1827). Also in AGN, SHMH, Dirección General de Minería, OL 164, OL 175, and OL 186. See also Alcalde Mongrut, *Mariano de Rivero,* 29–39.

15. Hunt, *Price and Quantum Estimates.*

16. Rivero y Ustáriz, *Colección de memorias;* Fischer, *Minas y mineros.*

17. Hunt, *Price and Quantum Estimates.*

18. Fisher, *Minas y mineros.* See also John R. Fisher, ed., "Matrícula de los mineros del Perú, 1790" (Lima: Seminario de Historia Rural Andina, Universidad Nacional Mayor de San Marcos, 1975) (mimeo); and Fisher, "Mineros y minería de plata en el virreinato del Perú, 1776–1824," *Histórica,* 3, 2 (Pontificia Universidad Católica del Perú, Lima, 1979): 57–70. See also Javier Tord and Carlos Lazo, *Hacienda, comercio, fiscalidad, y luchas sociales (Perú colonial)* (Lima: Biblioteca Peruana de Historia, Economía y Sociedad, 1981), 204–6, 207 (table).

19. On the war with Chile see Jorge Basadre, *Chile, Perú, y Bolivia independientes* (Barcelona: Salvat Editores, 1948); Basadre, *Historia de la república,* particularly vol. 8; Nelson Manrique, *Las guerrillas indígenas en la guerra con Chile* (Lima: Centro de Investigación y Capacitación, 1981); Mallon, *Defense of Community,* 80–122; Heraclio Bonilla, "The War of the Pacific and the National and Colonial Problem in Peru," *Past and Present,* no. 81 (November 1978): 92–118; and, more recently, Gastón Antonio Zapata, "La crise de l'état national au Pérou pendant la guerre du Pacifique, 1879–1883." (Thèse d'histoire, Ecole des Hautes Etudes en Sciences Sociales, Paris, 1986). See also the commentary of Manrique on Basadre's work in Nelson Manrique, "Basadre y la guerra del Pacífico," in *Jorge Basadre: La política y la historia,* ed. Noé Jave (Lima: Lluvia Editores, 1981), 191–225.

20. I refer to the Braudelian distinction between *court, moyenne,* and *longue durée* (short, medium, and long time spans). See Fernand Braudel, *The Mediterranean and the Mediterranean World in the Age of Philip II,* trans. from the French by Siân Reynolds (New York: Harper, 1976; original edition, 1949). See also Braudel, *Ecrits sur l'histoire,* 9–83, a discussion of *les temps de l'histoire* (the times of history). Finally, see also Ernest Labrousse, *Esquisse du mouvement des prix et des revenus en France au dix-huitième siècle* (Paris: Dalloz, 1933).

21. Rosemary Thorp and Geoffrey Bertram emphasize also the decade of 1890 as a transition period in mining to a more "open economy." Thorp and Bertram, *Peru 1890–1977: Growth and Policy in an Open Economy* (London: Macmillan, 1978), 72–95.

22. The Peruvian peso was exchanged for the British pound sterling at the rate of one pound to five pesos until the 1870s, and it was also equal to the American dollar. See Deustua, "Mines, monnaie," 2:634–745; Camprubí, *Historia de los bancos;* Eduardo Dargent C. *El billete en el Perú* (Lima: Banco Central de Reserva del Perú, 1979).

23. On guano, see again, Basadre, *Historia de la república,* particularly 3:147–65, and, in general, vols. 4–7. More specific are Shane Hunt, "Guano y crecimiento en el Perú del siglo diecinueve," *Hisla,* no. 4 (Revista Latinoamericana de Historia Económica y Social, Lima, 1984): 35–92 (the new edition in Spanish of his celebrated 1973 piece); and Bonilla, *Guano y burguesía.* On the diversion of funds from mining to guano production see Deustua, *Minería peruana,* 145–47.

24. In the 1890s Bolivia continued to rank third in world silver production. See Escuela Especial de Ingenieros de Lima, "Las minas de Bolivia," *Boletín de minas, industrias, y construcciones* 6 (Lima, 1890). See also Herbert Klein, *Bolivia,* 149–87.

25. Jiménez, "Reseña histórica," 50; Macera, *Estadísticas históricas,* 104.

26. Hunt, *Price and Quantum Estimates,* table 21, 57–59.

27. "Informe de la Aduana del Estado de Lima del 17 de setiembre de 1821," in AGN, SHMH, OL 10, caja 2. And in the twentieth century (1907), gold was still exported as coins from the port of Callao. See Biblioteca Nacional de Lima (hereafter BN), Serie Manuscritos Republicanos, E 1148/1907. Gold exports in powder from the port of Arica in 1869 are shown in PRO, FO 61. Consular Correspondence, Peru, vol. 260 (1870). Mr. Jerningham, Commercial, f. 477.

28. For figures on world gold production , see Vilar, *Or et monnaie dans l'histoire* (Paris: Flammarion, 1974), annexe 2, 431–33. In 1900, when there are already reliable figures on national production, Peru produced 1,633 kilograms, 0.47 percent of the world's output (349,130.25 kg.).

29. See Fisher, ed., "Matrícula de los mineros," and compare "Razón de la matrícula general de los mineros de 1790" with "Estado general de minería" of 1799; see also Deustua, "El ciclo interno de la producción del oro en el tránsito de la economía colonial a la republicana: Perú, 1800–1840," *Hisla,* no. 3 (Revista Latinoamericana de Historia Económica y Social,

Lima, 1984), particularly 28–29; and finally AGN, SHMH, PL 6, no. 193–194, "La Municipalidad de Ayacucho a nombre de los vecinos de ese departamento sobre que no se adjudiquen al estado las minas que expresa."

30. See AGN, SHMH, Prefectura de La Libertad, OL 131, caja 38, ff. 398–488 (1825); and also "Informe de la Diputación Territorial de Minería de Pataz," AGN, Serie Minería C-12, legajo 72, Correspondencia (1846).

31. "Informe del subprefecto de Carabaya, Pablo Pimentel, al prefecto de Puno y al ministro de hacienda, J. F. Melgar," BN, Manuscritos Republicanos, D 1696 (1849). See also "Informe del Tribunal de Minería a la Diputación de Puno sobre la fijación de los descubrimientos de oro en Carabaya," AGN, Libros Manuscritos Republicanos H-4–0450 (1849).

32. AGN, SHMH, Prefectura de La Libertad, OL 131, caja 38, ff. 398–488 (1825).

33. "Propuesta de la junta de mineros de Huallanca," AGN, Serie Minería C-12, legajo 71 (1828); "Informe de la Diputación Territorial de Minería de Pataz," legajo 72, Correspondencia (1846).

34. On this regard see also Deustua, "Ciclo interno," 25–28.

35. In Heraclio Bonilla, ed., *Gran Bretaña y el Perú: Informes de los cónsules británicos, 1828–1919,* 4 vols. (Lima: Instituto de Estudios Peruanos and Fondo del Libro del Banco Industrial, 1975), 1:201–30. For criticism and use of British consular and trade reports see Gootenberg, *Between Silver and Guano,* particularly 12–14 and 18–20. Other printed sources containing British and American consular reports are R. A. Humphreys, ed., *British Consular Reports on the Trade and Politics of Latin America, 1824–1826* (London: Royal Historical Society, 1940); and William R. Manning, ed., *Diplomatic Correspondence of the United States Concerning the Independence of the Latin American Nations,* vol. 3 (New York: Oxford University Press, 1925); and *Diplomatic Correspondence of the United States: Inter-American Affairs, 1831–1860* (Washington D.C.: Carnegie Endowment for International Peace, 1938). British consular reports, of course, are in PRO, London, FO; particularly the printed ones in the *Accounts and Papers* of the *British Parliamentary Papers.*

36. Nils Jacobsen, "Landtenure and Society in the Peruvian Altiplano: Azángaro province, 1770–1920" (Ph.D. dissertation, University of California, Berkeley, 1982), 321–22. See also Jacobsen, *Mirages of Transition,* 185.

37. Teodorico Olaechea, *Apuntes sobre la minería en el Perú* (Lima: Imprenta de la Escuela de Ingenieros, 1898), 19.

38. See also Thorp and Bertram, *Peru 1890–1977,* 76 (table 5.2).

39. I have calculated gold prices in British pounds sterling, shillings, and pennies (based on the sources for silver prices that I used for fig. 2.2), as well as the value of gold in relation to silver. See also Adolph Soetbeer, *Edelmet-all-Produktion* (Gotha: Justus Perthes, 1879), 130–31; Vilar, *Or et monnaie*, annexe 2, 431–33; Y. S. Leong, *Silver: An Analysis of Factors Affecting Its Price* (Washington D.C.: Brookings Institution, 1933), 5; and Mitre, *Patriarcas*, 30 (table 3), 35 (table 6).

40. Vilar, *Or et monnaie*, 397–409; Paul Bairoch, *Commerce extérieur et développement économique de l'Europe au dix-neuvième siècle* (Paris: Mouton and Ecole des Hautes Etudes en Sciences Sociales, 1976), 36–38.

41. To obtain the value of gold production, I have multiplied production figures (amount) in those years by its prices. The sources for both are mentioned in fig. 2.3 and notes 25 and 39 in this chapter. For 1897 I have used Olaechea's figures.

42. Heraclio Bonilla, "Aspects de l'histoire économique et sociale du Pérou au dix-neuvième siècle," 2 vols. (thèse de doctorat du troisième cycle, Université de la Sorbonne, Paris, 1970); Bonilla, "Coyuntura comercial," tables 5, 7, 8.

43. Bonilla's sources were mostly accounts of French and British imports. The first piece he published on this theme was a paper presented at the 38th Americanists International Congress in Stuttgart in August 1968. This paper was later published as an article in *Revista del museo nacional* 35 (Lima, 1967–68; actually printed in 1970). His 1970 doctoral thesis contains this original piece. I have worked mostly with the data in this original article, which was later reproduced without its graphics, in Heraclio Bonilla, *Un siglo a la deriva: Ensayos sobre el Perú, Bolivia, y la guerra* (Lima: Instituto de Estudios Peruanos, 1980), 13–46.

44. Hunt, *Price and Quantum Estimates*, 69, 37.

45. Ibid., 69 (table 25).

46. Bolivia until the early 1840s used the Peruvian port of Arica for its exports. See the 1826 report of the British consul in Peru, Charles Milner Ricketts, in Bonilla, ed., *Gran Bretaña y el Perú*, 1:27; see also Bonilla, *Siglo a la deriva*, 17. Bolivia and Chile also produced copper. Toward 1850, for example, the Bolivian mining center of Coro Coro was producing some 50,000 quintals of copper per year (Tristan Platt, personal communication). Chilean copper production can be seen in A. Herrmann, *La producción en Chile de los metales i minerales desde la conquista hasta fines del año 1902* (Santiago: n.p., 1903); Marcello Carmagnani, *Les mécanismes de la vie*

économique dans une société coloniale: Le Chili, 1680–1830 (Paris: SEVPEN, 1973); Vayssière, *Siècle de capitalisme.*

47. Mario Samamé Boggio, *El Perú minero* (Lima: Instituto Geológico Minero y Metalúrgico, 1981), 5:287–313, 327–52.

48. Bonilla, "Coyuntura comercial"; also Bonilla, *Aspects de l'histoire;* and Bonilla, *Los mecanismos de un control económico,* vol. 5 of Gran Bretaña y el Perú (Lima: Instituto de Estudios Peruanos and Fondo del Libro del Banco Industrial, 1977).

49. Regulus is the impure metal produced by the smelting of ore. Bonilla, *Mecanismos de control,* tables 12, 13, 14.

50. AGN, SHMH, OL 10, caja 2, f. 2 (1821); and OL 40, caja 6, ff. 96–97 (1821).

51. BN, Manuscritos Republicanos, D 6772 (1826).

52. Rivero y Ustáriz, *Colección de memorias,* 2:20–21.

53. AGN, Sección Casa de Moneda, Expedientes CMR 0095, CMR 0098, CMR 0099, CMR 00100, and CMR 00101 (1828, 1829).

54. "Informe de la Aduana del Estado de Lima del 17 de setiembre de 1821," AGN, SHMH, OL 10, caja 2, f. 27 (1821). See also Sergio Villalobos, *Comercio y contrabando en el Rio de la Plata y Chile* (Buenos Aires: Editorial Universitaria, 1965), 11. Villalobos mentions that by the eighteenth century merchants from Chile were already selling copper in the French ports of Bordeaux and Marseilles.

55. On kitchen and dining utensils made of silver, see Oficialía Mayor de Cultura, *Platería civil* (La Paz: Museos Municipales, 1992).

56. AGN, SHMH, OL 10, caja 2, f. 14 (1821).

57. The same was said about France during the Old Regime, where "nos campagnes étaient saturées d'industrie" (our rural areas were full of industry). In other words, there was no clear separation between agriculture and the production of industrial goods and crafts made locally by peasants. Emile Coornaert, preface to Jean Vidalenc, *La petite métallurgie rurale en Haute-Normandie sous l'ancien régime,* cited in Franklin Mendels, "Des industries rurales à la protoindustrialisation: Historique d'un changement de perspective," *Annales: Economies, sociétés, civilizations* 39, 5 (September–October 1984): 977.

58. AGN, Sección Casa de Moneda, Expedientes CMR 0095, CMR 0098, CMR 0099, CMR 00100, and CMR 00101 (1828, 1829). See also PRO, FO, "Foreign and Colonial Merchandize Imported into the United Kingdom from Peru," in *Accounts and Papers* of the *British Parliamentary Papers,* vol. 39 (1849), p. 372; and Bonilla, *Mecanismos de control,* 175–80.

59. See Oficialía Mayor de Cultura, *Platería civil.*

60. In 1846, for example, Carlos Renardo Pflucker wrote that "copper minerals, which are abundant in this country [Peru] and are exported as raw or refined minerals, form from a long time, one of the main basis of the powerful and flourishing trade of one of our neighboring Republics [Chile]." Pflucker: *Exposición que presenta al supremo gobierno con motivo de las ultimas ocurrencias acaecidas en la hacienda mineral de Morococha* (Lima: Imprenta del Correo Peruano, 1846), 7; see also Carmagnani, *Mécanismes de la vie,* 208 (table 2).

61. AGN, SHMH, OL 10, caja 2, f. 2 (1821).

62. AGN, SHMH, Tesorería Departamental, OL 40, caja 6, ff. 96-97 (1821).

63. Rivero y Ustáriz, *Colección de memorias,* 1:196-97, 2:5.

64. See Bonilla, "Coyuntura comercial," table 7.

65. Jiménez, "Reseña histórica," 48. The price of £125 per ton was equal to 28 pesos per quintal, 6 pesos (or 27%) above the price of copper in 1821.

66. Ibid.

67. Archivo de la Universidad Nacional de Ingeniería, Lima (hereafter AUNI), "Memoria del primer año" (thesis no. 2, alumno Segundo Carrión, Lima, 1878), p. 6.

68. AUNI, "Memoria de Viage," (thesis no. 1, alumno Pedro Félix Remy, Lima, 1878).

69. AGN, "Minuta de constitución de la sociedad Garland y Compañía," Lima, April 1, 1879, Protocolo notarial no. 591, ff. 742-43 (escribano Francisco Palacios).

70. Pflucker, *Exposición,* 7.

71. Ibid., 7-12.

72. AGN, Serie Minería C-12, legajo 76 (1846). On the importation of German workers to Tucto, Morococha, due to the efforts of Leonardo Pflucker (who by that time was studying mineralogy in Germany), see also Flores Galindo, *Mineros de la Cerro de Pasco,* 37.

73. Dirección de Estadística, *Estadística de las minas de la república del Perú en 1878* (Lima: Imprenta del Estado, 1879), 195-97.

74. AGN, Serie Minería C-12, legajo 74 (1833).

75. Rivero y Ustáriz, *Colección de memorias,* 2:2, 3, 20.

76. "Informe de la Aduana del Estado de Lima del 17 de setiembre de 1821," AGN, SHMH, OL 10, caja 2, f. 27 (1821).

77. On the Bourbon reforms see John Lynch, *Bourbon Spain, 1700-1808* (Oxford: Basil Blackwell, 1989), particularly 329-74. See also D. A. Brading,

"Bourbon Spain and its American Empire," in *Colonial Spanish America,* ed. Leslie Bethell (Cambridge: Cambridge University Press, 1988), 112–62. Specifically on the impact of the Bourbon reforms on mining in colonial Peru, although focusing exclusively on silver mining, see Fisher, *Minas y mineros;* see also Carlos Deustua Pimentel, "La minería peruana en el siglo dieciocho (aspectos de un estudio entre 1790 y 1796)," *Humanidades 3* (Pontificia Universidad Católica del Perú, Lima, 1969): 29–47. On Bolivia, see Enrique Tandeter, *Coacción y mercado: La minería de la plata en el Potosí colonial, 1691–1826* (Cusco: Centro de Estudios Regionales Andinos Bartolomé de las Casas, 1992, 209–68; original edition, in Spanish, of *Coercion and Market*); on Mexico, David A. Brading, *Miners and Merchants in Bourbon Mexico* (Cambridge: Cambridge University Press, 1970).

78. Bonilla, *Mecanismos de control.*

79. Hunt, *Price and Quantum Estimates.*

80. The document is the "Foreign and Colonial Merchandise Imported into the United Kingdom from Peru," in the Foreign Office, London. See references in the following graph. The figures are similar to Bonilla, *Mecanismos de control,* 193 (table 20), except for minimal differences in some years.

81. On this regard see Landes, *Unbound Prometheus,* particularly 88–100, 193–95.

82. Hunt, *Price and Quantum Estimates,* 63 (table 23).

83. Bonilla, "Coyuntura comercial," table 7.

84. Mitre, *Patriarcas,* particularly 180–93; Klein, *Bolivia,* 149–87; Augusto Guzmán: *Historia de Bolivia* (Cochabamba: Editorial Los Amigos del Libro, 1990), 248–54.

85. See Erick D. Langer, *Economic Change and Rural Resistance in Southern Bolivia, 1880–1930* (Stanford: Stanford University Press, 1989), particularly 24–27; Fernando Cajías, *La provincia de Atacama, 1825–1842* (La Paz, 1975); and, on the strategic role of the Bolivian city of Tarija, and its merchants, as an entrepôt "between the southern Bolivian silver mines, the eastern Chaco frontier, the Pacific coast, and northwestern Argentina," see Erick D. Langer and Gina L. Hames, "Commerce and Credit on the Periphery: Tarija Merchants, 1830–1914," *Hispanic American Historical Review* 74, 2 (May 1994): 288.

86. Bonilla, *Siglo a la deriva,* 17. A particular Peruvian and Bolivian tree bark was exported to extract quinine from it.

87. "Comunicación de Juan Antonio Gordillo, Administrador General de la Aduana del Estado de Lima, al Excelentísimo Señor Don José de San

Martín, Capitán General y Protector del Perú" (29 October 1821), AGN, SHMH, OL 10, caja 2, f. 41 (1821).

88. On the socioeconomic unity between the southern Peruvian Andes and Bolivia see Glave, *Trajinantes,* for the colonial exchanges and integration from Potosí to Cusco; on an Aymara ethnic nation in southern Peru, see Glave, *Vida, símbolos, y batallas: Creación y recreación de la comunidad indígena: Cusco, siglos dieciséis-veinte* (Mexico City: Fondo de Cultura Económica, 1992); and Alberto Flores Galindo, *Arequipa y el sur andino, siglos dieciocho-veinte* (Lima: Editorial Horizonte, 1977), particularly 45-61. Peru and Bolivia were unified as a single nation during the years of the Peruvian-Bolivian Confederation (1836-39).

89. Obviously, as with copper, it was preferable usually to export refined tin rather than tin ore, although that was not always the case. The raw ore was more available than the refined metal, which would suppose a previous investment of capital to set the refining process.

90. See Agustín Telles, "Método que siguen en los trapiches los pucheros cagchas" (1831), documentary appendix of Tristan Platt, "The Ayllus of Lipez in the Nineteenth Century," paper presented at the 44th Americanists International Congress, Manchester, September 1982, pp. 125-27.

91. Oscar Alayza, *La industria minera en el Perú, 1936* (Lima: Ministerio de Fomento, 1937), 108.

92. Samamé Boggio, *Perú minero,* 5:340, 360.

93. According to Samamé Boggio, for example, the Quechua language in the country differentiates between *anta* (copper), *chumpe* (bronze), and *ttiti* (tin). Samamé Boggio, *Perú minero,* 5:359. See also Roger Ravines, ed., *Tecnología andina* (Lima: Instituto de Estudios Peruanos and Instituto de Investigación Tecnológica Industrial y de Normas Técnicas, 1978), 475-554.

94. This figure is estimated in pesos of 8 reales each, pesos of the more stable monetary period of 1830-1870. See Deustua, "Mines, monnaie," 2:634-745; Camprubí, *Historia de los bancos;* Dargent, *Billete en el Perú.* See also José Deustua, "De la minería a la acuñación de moneda y el sistema monetario en el Perú del siglo diecinueve," in *Apuntes sobre el Proceso Histórico de la Moneda. Perú, 1820-1920,* ed. Javier Ramírez Gastón and Soledad Arispe (Lima: Banco Central de Reserva del Perú, 1993), 79-140.

95. Calculations are based on table 2.1, and the data of figs. 2.1, 2.2, 2.3, 2.4, 2.5, and their sources.

96. See chapters 3 and 4 in this book; see also Deustua, *Minería peruana,* 111-220; and "Mines, Monnaie," chaps. 4 and 5.

97. See, for example, Janet Campbell Ballantyne, "The Political Economy of Peruvian 'Gran Minería'" (Ph.D. dissertation, Cornell University, 1976); Elizabeth Dore, "Accumulation and Crisis in the Peruvian Mining Industry, 1900-1977" (Ph.D. dissertation, Columbia University, 1980); ECO, *Crisis minera y sobre-explotación de la fuerza de trabajo* (Lima: ECO, Grupo de Investigaciones Económicas, 1980). On the development of the iron and steel industries and the political and trade-union role of their workers, see also Martha Rodríguez Achung, "Interpretación de la historia político-sindical del proletariado siderúrgico, 1957-1972" (Lima: Pontificia Universidad Católica del Perú, Taller de Estudios Urbano Industriales, 1980) (mimeo).

98. Jonathan Levin, *The Export Economies: Their Pattern of Development in Historical Perspective* (Cambridge, Mass.: Harvard University Press, 1960); José Manuel Rodríguez, *Estudios económicos y financieros y ojeada sobre la hacienda pública del Perú y la necesidad de su reforma* (Lima: Librería Gil, 1895), particularly 317-18.

99. Bonilla, *Guano y burguesía;* Hunt, "Guano y crecimiento," particularly 53 and note 39; and Javier Tantalean Arbulú, *Política económico-financiera y la formación del estado, siglo diecinueve* (Lima: Centro de Estudios para el Desarrollo y la Participación, 1983), particularly 68-75. More recently, Gootenberg agrees with Hunt's estimates; see Gootenberg, *Imagining Development,* 2.

100. The official basic Peruvian currency until 1863 was the peso and was going to be equal to the sol introduced thereafter until well into the twentieth century. But the sol was going to experience a devaluation crisis since 1873. The peso and the stable sol were exchanged on equal terms with the U.S. dollar and at the rate of five pesos or soles to the British pound sterling. I also assume in these calculations, then, a stable currency, whether pesos, soles, or dollars. On the Peruvian monetary system in the nineteenth century, see Deustua, "Mines et monnaie," 2:634-745, and "Minería a la acuñación"; Camprubí, *Historia de los bancos;* Dargent, *Billete en el Perú.*

101. With the monetary reform of 1901, the exchange rate was ten soles to the British pound sterling.

102. On the continuation of the guano industry after the boom years and its collapse, see Pablo Macera, "Historia de la compañía administradora del guano" (Lima, 1968 [mimeo]), reprinted as "El guano y la agricultura peruana de exportación, 1909-1945," in Macera, *Trabajos de historia,* 4:309-499.

103. This same argument is made to measure their contribution to the Peruvian internal market in José Deustua, "Routes, Roads, and Silver Trade in Cerro de Pasco, 1820–1860: The Internal Market in Nineteenth-Century Peru," *Hispanic American Historical Review* 74, 1 (1994): 1–31, particularly 4–5. See also chapter 4 of this book.

104. See Mitre, *Patriarcas* and "Economic and Social Structure of Silver Mining in Nineteenth-Century Bolivia" (Ph.D. dissertation, Columbia University, 1977); Klein, *Bolivia*, chap. 6; Vayssière, *Siècle de capitalisme*.

105. As early as 1898, however, Manuel Clotet, an adviser and friend of the Fernandini mining company, admonishing Eulogio Fernandini: "Adapt, I have told you. These [scraps of silver] amount to nothing but a waste of time, money, and patience, all the more so now that you inform me you have sold your smelted bars at 7.80 soles. . . . It would not be surprising if the bronzes of Colquijirca contain copper. This is the best business of the day." Clotet to Fernandini, 23 April 1898, Lima, AFA, Serie Algolán, ALG 205.

Chapter 3

1. All the data I present in this chapter will be in varas. A Castilian vara measured 0.836 meters, ten varas then were 8.36 meters and one vara and a half, 1.254 meters. About the mining legislation, apart from the sources mentioned in note 2 below, see Eduardo García Calderón, *Constituciones, códigos, y leyes,* in two volumes, kept in the Library of the Archive of the Ministry of Foreign Relations (Archivo del Ministerio de Relaciones Exteriores, Lima; hereafter AMRE). Some discussion of nineteenth-century mining laws can be also seen in BN, Sección Manuscritos Republicanos (hereafter SMR), expediente E 929/1925, which includes a typed text of some 110 folios; see also BN, SMR, E 992/1929.

2. For the legal background to nineteenth-century Peruvian mining, see Ricardo Arana, *Colección de leyes, decretos, y resoluciones que forman la legislación de minas del Perú* (Lima: n.p., n.d.), which contains the three most important legal documents of nineteenth- and early-twentieth-century Peruvian mining: the Nuevas Ordenanzas de Minería of 1786; the Ley de Reforma de la Minería of 1877, which was approved just one year after the law of creation of the Escuela de Ingenieros del Perú (Peruvian school of mining); and the Código de Minería of 1900, which definitively abolished the previous two. See also Juan Crissóstomo Nieto and Mariano Santos de

Quirós, *Colección de leyes, decretos, y ordenes publicadas en el Perú desde su independencia* (Lima: Imprenta de la Colección, 1864); Francisco García Calderón, *Diccionario de la legislación peruana,* 2d ed. (Paris: Librería de Laroque, 1879); Eduardo Habich, "Código de minería," in *Anales de la Escuela de Construcciones Civiles y de Minas del Perú,* vol. 3 (Lima, 1883). Finally, the new *Código de minería* of 1900 was also published with comments in Paulino Fuentes Castro, ed., *Nueva legislación peruana* (Lima: Editor de El Diario Judicial, 1903). The Nieto and Santos de Quirós, *Colección de Leyes,* is well kept in the library of AMRE. It is composed of 13 volumes, with 2 volumes of indexes. I consulted this collection in AMRE in April 1982.

3. On the general legal history of the nineteenth century, still an undeveloped field of Peruvian historiography, see Fernando de Trazegnies, *La idea del derecho en el Perú republicano del siglo diecinueve* (Lima: Pontificia Universidad Católica del Perú, 1980); and Carlos Augusto Ramos, *Toribio Pacheco: Jurista peruano del siglo diecinueve* (Lima: Pontificia Universidad Católica del Perú, 1993).

4. I have omitted, obviously, twentieth-century statistics kept in the Padrones Generales de Minas, several of which are in the Archivo General de la Nación, Lima (hereafter AGN). Pablo Chaca, in *Capitalismo minero* (Lima: Universidad Nacional Mayor de San Marcos, 1980) has used this kind of material, as has Tarnawiecki in "Crisis y desnacionalización." The *Padrones generales de minas* from 1899–1902 to 1920 are in AGN, Serie Impresos H-6, H-6–0718 to H-6–0729. There is also a *Registro oficial de fomento, minas, e industria* from 1901 to 1906 in H-6–0752 to H–6–0776. Finally the *Extractos estadísticos del Perú* for 1920, 1924, 1925, and 1929–1930 are in H-6–0410 to H-6–0413. These quantitative sources are extremely important to the study of Peruvian mining in the early twentieth century.

5. In 1822, J. García del Rio and Diego Paroissien, in a "Memoria sobre el estado del Perú" addressed to the British government (London, 5 November 1822), repeated the accounts on the number of mines contained in the *matrícula* of 1790. According to this report, "at the end of the last century in the Provinces of Peru there were 670 mines at work *[minas en labor]*, and 578 mines not working *[paradas]*, without taking into account the places where gold deposits are washed *[lavaderos]* and the azogue mines, particularly those of Huancavelica." See Colección Documental de la Independencia del Perú, *Misión García del Rio-Paroissien* (Lima: CDIP, 1973), tome. 11, vol. 2, 76–77.

6. The first modern census was taken in 1876. The only other equally general census before that was that of Viceroy Francisco Gil de Taboada of

1791–1795. Both population records were taken in the periods of the mining statistics I am working with, the decades of 1790 and 1870, during which mining statistics reveal new interest in exploration and development. Both moments of Peruvian history, then, were significant periods of scientific and statistical experimentation. One evaluation of the 1876 census is that of Alida Díaz, "El censo general de 1876 en el Perú," Lima, Seminario de Historia Rural Andina, Universidad Nacional Mayor de San Marcos, 1974 (mimeo). On the scientific and statistical burgeoning of the 1870s, particularly the role in it of Manuel Atanasio Fuentes, see Gootenberg, *Imagining Development*, 64–71.

7. Dirección de Estadística, *Estadística de las minas*. Manuel A. Fuentes, responsible too for the population census of 1876, was the director of the statistics section (la Dirección de Estadística) of the Ministry of Promotion and Public Works (el Ministerio de Fomento y Obras Públicas). The table containing these data, however, was signed by B. Fonseca on 31 December 1878. The data presented here are from pp. 92–93 of the *Estadística*. I have found copies of the *Estadística* in BN and in the library of the Instituto Científico y Tecnológico del Perú, Lima.

8. *Boletín de minas, industria, y construcciones*, Escuela Especial de Ingenieros de Lima, year 3, vol. 3, Lima, 1887.

9. "Informe de José Manuel Sorogastúa al Tribunal General de Minería del 23 de enero de 1833," AGN, Serie Minería C-12, legajo 74 (1833).

10. See José Ignacio López Soria, "La Escuela de Ingenieros y la minería," in *Historia, problema, y promesa: Homenaje a Jorge Basadre,* ed. Francisco Miró Quesada, et al., 2 vols. (Lima: Pontificia Universidad Católica del Perú, 1978), 2:149–69. See also José Ignacio López Soria, *Universidad Nacional de Ingeniería: Los años fundacionales, 1876–1909* (Lima: Universidad Nacional de Ingeniería, 1981).

11. A good contemporary dictionary of technical and mineral names is "Diccionario de algunas voces técnicas de mineralogía, y metalurgia municipales en las mas provincias de este Reyno del Perú, indicadas por orden alfabético y compiladas por los autores del mismo *Mercurio,*" *Mercurio peruano* 1 (January 1791): 73–89 (facsimile reproduction of the original in the Biblioteca Nacional, Lima). A much older dictionary is that of Garcia Llanos, *Diccionario y maneras de hablar que se usan en las minas y sus labores en los ingenios y beneficios de los metales,* ed. Thierry Saignes and Gunnar Mendoza (La Paz: Museo Nacional de Etnografía y Folklore, 1983; original edition, n.p., 1609.). See also Frédérique Langue and Carmen Salazar-Soler, *Dictionnaire des termes miniers en usage en Amérique espagnole*

(seizième–dix-neuvième siècle) (Paris: Editions Recherche sur les Civilisations, 1993).

12. On the oil question during colonial times, see Pablo Macera, "Las breas coloniales del siglo dieciocho," in *Trabajos de historia*, 3:229–74; and for the early twentieth century, see Dilma Dávila, "Talara, los petroleros, y la huelga de 1931" (tesis de sociología, Programa Académico de Ciencias Sociales, Pontificia Universidad Católica del Perú, Lima, 1976). See also BN, SMR, E 479 (1917), which deals with the oil workers' strikes of November and December 1917 in the oil fields of Talara, Negritos, and Lobitos; and BN, SMR, E 1148 (1907), which deals with oil exports and profits. While guano exports resulted in some £150,000 in profits per year toward 1907, three British companies reported the following profits from their Peruvian oil exports: Lobitos Oil Fields, £158,000; Lagunitas, £22,000; and London and Pacific Petroleum Company, £340,000. The profit from the export of each ton of petroleum was estimated to be £2. See also Alejandro Garland, *La industria del petróleo en el Perú en 1901*, Boletín no. 2, Cuerpo de Ingenieros de Minas, Lima, 1902.

13. "In the previous years of the war declaration the fiscal revenues were pledged, the liberalities of the Government quite arrested, and the import payments unbalanced, all these due to the breakdown of the guano returns, which paid the exports difference. Many people who nurtured from the Government were dedicated then to obtain metals even from scraps, and to put them in domestic trade at good prices because they were highly demanded by merchants who needed metals to contain the rise of the paper money exchanges." Luis Esteves, *Apuntes para la historia económica del Perú* (Lima: Centro de Estudios de Población y Desarrollo, 1971; original edition, 1882), 83. See also Camprubí, *Historia de los bancos*, 316–416.

14. See, for example, Mallon, *Defense of Community*, 80–122; and Manrique, *Guerrillas indígenas*.

15. On this and on the evolution of the Lima School of Engineers, later the University of Engineering, see López Soria, *Historia de la Universidad Nacional de Ingeniería*, particularly 71–152.

16. See Deustua, "Mines et Monnaie," 2:634–745; Deustua, "Minería a la acuñación"; Camprubí, *Historia de los bancos;* Dargent, *Billete en el Perú.*

17. "Razon de la matrícula general de los mineros" and "Estado general de minería" of 1790 and 1799, both in Fisher, ed., "Matrícula de los mineros," 33, 34.

18. AGN, SHMH, OL 131, caja 38, ff. 489–639 (1825).

19. Ibid.

20. "Lista or matrícula de los operarios de minas y hasiendas . . . ," AGN, Serie Minería C-12, legajo 61 (1827).

21. AGN, SHMH, OL 225, ff. 569–629 (1833).

22. Fisher, ed., "Matrícula de los mineros," 14–19, 22–26.

23. The average of 38 percent is based on the three mining centers on which I have data: Pasco (42%), Hualgayoc (30%), and Huallanca (41%).

24. Deustua, *Minería peruana,* esp. 35.

25. Fisher, ed., "Matrícula de los mineros," 14–18; and "Lista or matrícula de los operarios de minas y hasiendas," AGN, Serie Minería C-12, legajo 61 (1827).

26. AGN, testamento de José Lago y Lemus, escribano José de Selaya, 16 January 1838 and 17 January 1838, Lima, Protocolos Notariales n. 700, ff. 7–9r.

27. See . "Lista or Matrícula de Minas y Hasiendas . . . ," AGN, Serie Minería C-12, legajo 61 (1827); and *Padrón general de minas correspondiente al segundo semestre del año de 1878* (Lima: Imprenta del Estado, 1878), 6–11. See also AGN, Poderes de Pedro Abadía of 21 October 1807 and 11 August 1820, Lima, Protocolo Notarial 876, f. 98 and Protocolo Notarial 880, f. 56, Escribano Manuel Suárez; see also Deustua, *La minería peruana,* 125–30, and the discussion that follows further on in this chapter.

28. The existence and role of this mining elite have been regularly ignored by Peruvian historiography, although not by Florencia Mallon in "Minería y agricultura en la sierra central: Formación y trayectoria de una clase dirigente regional, 1830–1910," in "Lanas y capitalismo en los Andes centrales," ed. Florencia Mallon (Lima: Taller de Estudios Andinos, Universidad Nacional Agraria de La Molina, 1977 [mimeo]).

29. As mentioned in the previous note, there are very few social studies of the Peruvian upper classes during the nineteenth century, particularly of the mining elite. The emphasis has been concentrated on the guano traders or the landed elites. For a study on the guano elite and its links with the Civilist party, see Yepes del Castillo, *Perú,* particularly 66–91.

30. Rivero y Ustáriz, "Memorial sobre el rico mineral del Cerro de Pasco," in *Colección de Memorias,* I, 186.

31. AGN. Serie Tributos. Legajo n. 6, cuaderno n. 179 (1827). "Matrícula de los operarios de minas en actual trabajo de los mineros de este asiento de Yauli . . . ."

32. Archivo de la Dirección Regional de Minería de Huancayo (hereafter ADRMH), Registro Cívico del Distrito de Yauli (1883), pp. 13–24.

33. Dirección de Estadísticas, *Censo general de la república del Perú formado en 1876* (Lima: Dirección de Estadísticas, 1878). See the section on the department of Junín, province of Pasco. On the 1876 population census, see Centro de Estudios de Población y Desarrollo, *Informe demográfico del Perú* (Lima: CEPD, 1972); Hunt, *Growth and Guano;* Díaz, *Censo general;* Macera, *Trabajos de historia,* esp. vol. 4; and Clifford T. Smith, "Patterns of Urban and Regional Development in Peru on the Eve of the Pacific War," in *Region and Class in Modern Peruvian History,* ed. Rory Miller (Liverpool: Institute of Latin American Studies, University of Liverpool, 1987), 77–101.

34. Dirección de Estadísticas, *Censo general,* section concerning the department of Junín, province of Pasco, district of Cerro.

35. Nelson Manrique, "El desarrollo del mercado interior en la Sierra Central, 1830–1910," Taller de Estudios Andinos, Universidad Nacional Agraria, Lima, 1979 (mimeo), 3 (table 1); Salvador Barrantes and Nora Velarde, *El capital internacional en la sierra central* (Lima: Universidad Nacional Federico Villareal, 1983), 10 (table 1); Manrique, *Mercado interno y región: La sierra central, 1820–1930* (Lima: Centro de Estudios y Promoción del Desarrollo [DESCO], 1987), 35 (table 2). The source of the data for all these works is the 1876 census analyzed by Waldemar Espinoza Soriano, *Enciclopedia departamental de Junín* (Huancayo: Editor Enrique Chipoco Tovar, 1973), 339.

36. Pablo Macera, *Población rural en haciendas* (Lima: Seminario de Historia Rural Andina, Universidad Nacional Mayor de San Marcos, 1976).

37. See Wilfredo Kapsoli, *Los movimientos campesinos en Cerro de Pasco, 1880–1963* (Huancayo: Instituto de Estudios Andinos, 1975); and Gerardo Rénique, "Sociedad ganadera del centro: Pastores y sindicalización en una hacienda alto-andina," Taller de Estudios Andinos, Universidad Nacional Agraria de La Molina, Lima, 1977 (mimeo).

38. Manrique, *Mercado interno y región,* 34–38.

39. On Hualgayoc, see Lewis Taylor, "Main Trends in Agrarian Capitalist Development: Cajamarca, Peru, 1880–1976" (Ph.D. dissertation, University of Liverpool, 1980); see also Taylor, "Earning a Living in Hualgayoc, 1870–1900," in Miller, ed., *Region and Class,* 103–24.

40. Dirección de Estadísticas, *Censo general,* section concerning the department of Junín, province of Pasco, district of Cerro.

41. On this problem, see Alfredo Torero, *El Quechua y la historia social andina* (Lima: Universidad Ricardo Palma, 1974); and Enrique Carrión Or-

doñez, "Fuentes bibliográficas sobre los idiomas del Perú," *Humanidades* 5 (Pontificia Universidad Católica del Perú, Lima, 1972–1973): 113–29.

42. The accuracy of travelers' estimates is discussed in Carlos Contreras, "Minería y población en los Andes: Cerro de Pasco en el siglo diecinueve" (research report, Instituto de Estudios Peruanos, Lima, September 1984; manuscript), esp. 19–26.

43. See ADRMCP, "Libro copiador de notas desde 1832 hasta 1835," Correspondencia, "Comunicación de la diputación de minería del Cerro de Pasco al subprefecto de la provincia," 14 January 1835, ff. 85v–86, in which the mining deputy argues, "los mas de los operarios de minas son becinos de las quebradas inmediatas, ó de provincias estrañas que solo vienen á este mineral á lograr de las pequeñas bonanzas que se presentan en las minas, ó a jornalear por dinero para subenir á sus gastos necesarios, y penciones que sobre ellos gravitan" (most of the mine workers live in the neighboring valleys or in faraway provinces and only come to this mining center to benefit from the small booms of the mines, or to work for a cash wage in order to pay their living expenses and financial burdens that weigh upon them). See also Deustua, *Minería peruana,* 209–11. For the end of the century, see André DeWind, "Peasants Become Miners: The Evolution of Industrial Mining Systems in Peru" (Ph.D. dissertation, Columbia University, 1977), chaps. 1 and 2; and Mallon, *Defense of Community,* 187–205.

44. Tschudi, *Testimonio del Perú,* 258.

45. Contreras, *Mineros y campesinos,* second edition, 112–3, table 15 and map n. 2.

46. ADRMCP, "Libro copiador de notas," "Comunicación de la Diputación de Minería del Cerro de Pasco a la Subprefectura," 26 June 1835, f. 106.

47. Contreras, *Mineros y campesinos;* DeWind, "Peasants Become Miners"; Mallon, *Defense of Community.* See also Joan Martínez Alier, *Los Huacchilleros del Perú: Dos estudios de formaciones sociales agrarias* (Lima and Paris: Instituto de Estudios Peruanos and Ruedo Ibérico, 1973).

48. Mallon, *Defense of Community,* 13–243.

49. On Hualgayoc, see Taylor, "Agrarian Capitalist Development"; and Taylor, "Earning a Living."

50. Flores Galindo, *Mineros de la Cerro de Pasco,* particularly 76–93.

51. Fisher, ed., "Matrícula de los mineros," 34.

52. In 1879 the Dirección de Estadística of the Ministerio de Hacienda organized and published the *Estadística de las minas de la república del Perú,* based on mining reports gathered from the different prefects and local

authorities throughout the country. However, this same publication contains data gathered before 1878 (in the section "Datos anteriores a los recogidos en 1878"). I am working with these previous data because the data in the main body of the *Estadística* do not mention the number of workers in the different mining centers of the country. See Dirección de Estadística, *Estadística de las minas,* 95–154.

53. Ministerio de Hacienda y Comercio, *Extracto Estadístico del Perú* (Lima: Ministerio de Hacienda y Comercio, 1931–1933), 134. There were 21,480 mine workers in Peru in 1915; 32,321 in 1929; and 14,408 in 1933.

54. This percentage has been calculated from the data of 1790–1799. There were at that time 8,875 mine workers in the entire country; their families accounted for an additional 35,500 people. There were 717 mineowners, with 2,868 family members. The total for both groups, including their families, is 47,960 people, or 4.3 percent of the general population, according to the census of the Viceroy Francisco Gil de Taboada. See Fisher, ed., "Matrícula de los mineros"; and Fisher, *Government and Society in Colonial Peru: The Intendant System, 1784–1814* (London: Athlone Press, 1970), app. 2.

55. "Lista or matrícula de los operarios de minas y hasiendas . . . ," AGN, Serie Minería C-12, legajo 61 (1827).

56. According to the *Padrón,* mines were classified by registration number. Many new mines and the new mineowners mentioned before appear in the last part of the Cerro de Pasco section, proving their recent registry (particularly between registry numbers 885 and 1102). See *Padrón general de minas correspondiente al segundo semestre,* 9–10.

57. On the European immigration to Peru, see Esteves, *Apuntes para la historia,* 7–12; Juan De Arona, *La inmigración en el Perú: Monografía histórico-crítica* (Lima: Imprenta del Universo, 1891); Abraham Padilla Bendezú, "Historia de la inmigración en el Perú," in *Juan De Arona y la inmigración en el Perú,* ed. Jorge Guillermo Llosa (Lima: Academia Diplomática del Perú, 1971), 217–62; on the Asiatic immigration, see Watt Stewart, *Chinese Bondage in Peru: A History of the Chinese Coolie in Peru, 1849–1874* (Durham, N.C.: Duke University Press, 1951); Vilma Derpich, "Introducción al Estudio del Trabajador Coolie en el Perú del siglo diecinueve" (tesis de historia, Universidad Nacional Mayor de San Marcos, Lima, 1976); Humberto Rodríguez Pastor, "Los trabajadores chinos culíes en el Perú: Artículos históricos," Lima: n.p., 1977 (mimeo); C. Harvey Gardiner, *The Japanese and Peru, 1873–1973* (Albuquerque: University of New Mexico

Press, 1975); finally, H. E. Maude, *Slavers in Paradise: The Peruvian Slave Trade in Polynesia, 1862–1864* (Stanford: Stanford University Press, 1981).

58. However there are references that show that in 1878 Malpartida was experiencing economic troubles. From the 44 mines he possessed, ten "could be claimed by any miner because their taxes have not been paid" ("eran denunciables por no haberse pagado la contribución"). See Ministerio de Hacienda y Comercio, *Padrón general de minas de 1887* (Lima: Imprenta del Estado, 1887), 11.

59. As a comparison, on the importance of family ties for the Monterrey elite in Porfirian Mexico, see Alex M. Saragoza, *The Monterrey Elite and the Mexican State, 1880–1940* (Austin: University of Texas Press, 1988), particularly 75–76 and fig. 2.

60. According to Nelson Manrique, who has studied the structure of land ownership in the Mantaro valley in the nineteenth century, there was no distinction between economic and family relationships within a particular family clan. See Manrique, "Desarrollo del mercado," 48. The family organization implied economic as well as social and kinship ties.

61. Ministerio de Hacienda, *Padrón general de minas,* 9, which says, "Ignacio Rey por la casa Steel" (in representation of the Steel house) and further on "Jorge E. Steel por M. Gutiérrez."

62. See Archivo Legal de la Empresa Minera Centro-Min Perú (Legal Archive of the Mining Company Centro-Min Peru, hereafter ALCMP), Lima. "Legajo de la Constitución de la Sociedad Minera Backus & Johnston del Perú," n. 45, 24 December 1896; see also Luis Alberto Sánchez, *Historia de una industria peruana: Cervecería Backus y Johnston S.A.* (Lima: Editorial Científica, 1978).

63. Chayanov, *Theory of Peasant Economy.*

64. Alan Knight, "The Peculiarities of Mexican History: Mexico Compared to Latin America, 1821–1992," *Journal of Latin American Studies* 24 (quincentenary supplement, 1992): 99–144, in which Knight refers to David W. Walker's book *Kinship, Business, and Politics: The Martínez del Río Family in Mexico, 1824–1867* (Austin: Austin University Press, 1986).

65. See also Deustua, *Minería peruana,* 131–34.

66. In Huallanca in 1833 of sixteen mineowners, thirteen had mayordomos and one, Enrique Tracy, Ricardo Spray "y Compañía," had two managers and one mayordomo. See AGN, SHMH, OL 225, ff. 569–629 (1833).

67. Dirección de Estadística, *Estadística de las minas,* 95–154, "Datos anteriores a los recogidos en 1878."

68. AUNI. The first file is that of Pedro Félix Remy, who studied at the School of Engineering between 1877 and 1880. He wrote two "memorias de viage [sic]" in 1878 to the "cerro mineral de Cajavilca (Ancash)" and the "mineral station" in Ica. Of particular interest are the reports of Segundo Carrión on Otuzco, Ica, Salpo, and Huamantanga (1878–79); Juan Garnier on Salpo, Otuzco, Ica, and Canta (1879); Federico Villareal on Yauli (1885); Ismael Bueno and Germán Remy on Cerro de Pasco (1887), and so on. For further reports, see below.

69. Celso Herrera and Felipe A. Coz, "Excursión a Huarochirí," AUNI, April 1889.

70. On the ethnic, racial, and cultural problem in nineteenth-century Peru see Bonilla, "War of the Pacific," 92–118; see also Françoise Morin, ed., *Indianité, ethnocide, indigenisme en Amérique latine* (Toulouse: Centre National de la Recherche Scientifique, 1982). On the social protest and the peasant movements in nineteenth-century Cerro de Pasco, see Kapsoli, *Movimientos campesinos,* particularly 72–73, 84–88.

71. On the chewing of coca leaves during the workday, see Carlos Y. Lisson, "Excursión a Parac," AUNI (1893), thesis 47. "Apart from this work [a working day], all the *carreteros* [those who pull the wagons] do *huarachi,* that is to say, work additionally one more day shift and rest one. In the moments of rest the workers chew coca leaves *[chacchan]*."

72. Ismael Bueno ("alumno"), "Informe sobre el distrito mineral de Yauli," AUNI, April 1885, file 14, studies by Ismael Bueno, 1882–1887.

73. Ibid.

74. Celso Herrera and Felipe A. Coz, "Excursión a Huarochirí," AUNI, April 1889.

75. On the enganche question see also Martínez Alier, *Huacchilleros del Perú;* Flores Galindo, *Mineros de la Cerro de Pasco;* Heraclio Bonilla, *El minero de los Andes: Una aproximación a su estudio* (Lima: Instituto de Estudios Peruanos, 1974); Daniel Cotlear, "Enganche, salarios y mercado de trabajo en la ceja de selva peruana," *Análisis* 7 (Cuadernos de Investigación, Lima, January–April 1979): 67–85; Cotlear, "El sistema de enganche a principios del siglo veinte: Una versión diferente" (tesis de economía, Pontificia Universidad Católica del Perú, Lima, 1979); and Mallon, *Defense of Community,* particularly 141–44, 223–25.

76. Cf. the example of the village of Muqui, cited by Flores Galindo in *Mineros de la Cerro de Pasco,* 120.

77. Some other original references to the enganche question can be found in Marco Aurelio Denegri, *La crisis del enganche* (Lima: San Martín y

Compañía, 1911); Pedro Zulen, "El enganche de indios," *La prensa* (Lima), 7 October 1910; and Hildebrando Castro Pozo, *Nuestra comunidad indígena* (Lima: Perugraph Editores, 1979; original edition, 1924), particularly 77-79, 81, 88-91.

78. Francisco R. del Castillo, "Excursión a Huarochirí y a Yauli," AUNI (1891), file 40 for 1892.

79. Ismael C. Bueno, "Asiento del Cerro de Pasco," *Boletín de minas, industria, y construcciones* (Escuela Especial de Ingenieros de Lima), year 3, vol. 3 (1887).

80. On the scorn of Indians and the campaigns to achieve their national vindication within Peruvian "indigenismo," see Luis Enrique Tord, *El indio en los ensayistas peruanos, 1848-1948* (Lima: Editoriales Unidas, 1978); José Tamayo Herrera, *Historia del indigenismo cuzqueño, siglos dieciseis-veinte* (Lima: Instituto Nacional de Cultura, 1980); and José Deustua and José Luis Rénique, *Intelectuales, indigenismo, y descentralismo en el Perú, 1897-1931* (Cusco: Centro de Estudio Rurales Andinos Bartolomé de las Casas, 1984).

81. Julio A. Morales, "Excursión a Huarochirí," AUNI, 11 June 1892, tesis no. 42 (1892).

82. Santiago Marrau, "Excursión a la provincia de Huarochirí," AUNI, tesis no. 48 (1894).

Chapter 4

1. Research on muleteering and internal trade in nineteenth-century Peru has been minimal until recently. Now, however, the bibliography is growing. See, for example, Magdalena Chocano, "Circuitos comerciales y auge minero en la sierra central a fines de la época colonial," *Allpanchis* 18, 21 (Instituto de Pastoral Andina, Cusco, 1983): 3-26 (an outgrowth of her previous work, "Comercio en Cerro de Pasco a fines de la época colonial" [tesis de historia, Pontificia Universidad Católica del Perú, Lima, 1982]); Nelson Manrique, "Los arrieros de la sierra central durante el siglo diecinueve," *Allpanchis* 18, 21 (Instituto de Pastoral Andina, Cusco, 1983): 27-46; Jaime Urrutia, "Comerciantes, arrieros, y viajeros huamanguinos, 1770-1870" (tesis de antropología, Universidad Nacional San Cristóbal de Huamanga, Ayacucho, 1982); and Urrutia, "De las rutas, ferias y circuitos en Huamanga," *Allpanchis* 18, 21 (Instituto de Pastoral Andina, Cusco, 1983): 47-64. See the entire issue of *Allpanchis* 18, 21, dedicated to *Arrieros y*

circuitos mercantiles andinos (Muleteers and Andean trade routes) (Cusco: Instituto de Pastoral Andina, 1983). This issue also contains a great etnographic description and analysis of the recent activity of arrieros and llameros in Huancavelica: Ricardo Valderrama and Carmen Escalante, "Arrieros, troperos, y llameros en Huancavelica," 65–88. Another present-day anthropological work on muleteering goes back in time using techniques of oral history: María Susana Cipolletti, "Llamas y mulas, trueque y venta: El testimonio de un arriero puneño," *Revista andina* 4 (Centro Bartolomé de las Casas, Cusco, 1984), 513–38. Finally, see Deustua, "Routes, Roads."

2. In addition to the references in note 1, see also Rodrigo Montoya, *Capitalismo y no capitalismo en el Perú: Un estudio histórico de su articulación en un eje regional* (Lima: Mosca Azul Editores, 1980), on commercial axis and economic circuits that were based in large parts on muleteering in late-nineteenth- and twentieth-century Peru, particularly in the regions of Pisco, Ica, Nazca, Puquio, and Abancay. On domestic and export trade, agrarian and mining production, and land tenure systems in the Azángaro area, Puno, from the eighteenth to the early twentieth centuries, see Jacobsen, *Mirages of Transition.*

3. "Informe del año 1897 sobre comercio y finanzas del Perú, por Alfred St. John," in Bonilla, ed., *Gran Bretaña y el Perú,* 1:277–96. The reference and quantitative data appear on pp. 282–83.

4. Ibid., 279. For the use of St. John's data to estimate the different regional amount of mineral transportation see Deustua, "Mines et monnaie," 1:225–9.

5. St. John, "Informe del año 1897," 279–80.

6. On the consumption of dependency theory, see Fernando Henrique Cardoso, "The Consumption of Dependency Theory in the United States," *Latin American Research Review* 12, 3 (1977): 7–24. For criticism of dependency theory at the time of its elaboration, see Ruggiero Romano, "Sous-développement économique et sous-développement culturel, à propos d'André Gunder Frank," *Cahiers Vilfredo Pareto* 24 (Geneva, 1971): 271–79; and Carlos Sempat Assadourian, "Modos de producción, capitalismo, y subdesarrollo en América Latina," in *Modos de producción en América Latina,* ed. Carlos Sempat Assadourian et al. (Cordoba: Cuadernos de Pasado y Presente no. 40, 1973).

7. See chapter 2. For a reference to different authors and estimations of guano revenues compared to those of silver, see also Deustua, "Routes,

NOTES TO PAGES 108-111

Roads," 5 (note 13). For the importance of copper exports between 1861 and 1864, see Bonilla, "Coyuntura comercial," table 7.

8. See, for example, "Certificados por los ensayadores de la Casa de Moneda sobre recepción de plata para su amonedación," BN, Sala de Investigaciones Bibliográficas, D 901, 1826; see also AGN, Sección Casa de Moneda, República, legs. 101, 102, 103, 104, CMR 00747, 00752–56, 00790–97, and 00830 (1843). Finally, see Deustua, "Routes, Roads," 10–14.

9. Apart from the sources in the previous note, see "Informe de José Abeleyra al Ministro de Hacienda," Archive of the National Museum of History, Lima (Archivo del Museo Nacional de Historia, hereafter AMNH), ms. 2082, July–December 1844; and Tschudi, *Testimonio del Perú,* 269–70.

10. In Deustua, "Routes, Roads" (esp. 8–11), I rather focused on José Abeleyra's report to the Ministry of Finances of 1844, in AMNH, ms. 2082. Changes between 1827 and 1844 (the dates of Landaburu's and Abeleyra's reports), however, were not really significant until the coming of the railroad, as we will see in the following pages. See also chapter 5.

11. Several proposals have been written on this topic. See Carlos Camprubí, *Bancos de rescate, 1821–1832* (Lima: n.p., 1963).

12. For further documentation on the debate over whether or not to establish bancos de rescate, see AGN, SHMH, PL 6, no. 114 (1826); OL 186, caja 117, ff. 652–61 (1829); and OL 216 (1832).

13. As late as 1869 a reporter for the Lima newspaper *El comercio* was commenting on how merchants in Cerro de Pasco "advance money to mineowners to buy raw silver" *(los comerciantes que les hacen algunos adelantos de dinero para piña).* The reporter, following the colonial custom, called the credit provider *aviador,* and—more significant—the mineowner was *su protegido* (his protegé), showing the firm grip that merchants still had on the economics of mining still at a very late date (1869). *El comercio* (Lima), 7 June 1869.

14. "Asuntos remitidos a la Comisión de Hacienda," Archive of the National Congress, Lima (Archivo del Congreso Nacional, hereafter ACNL), legajo 1, no. 16, "Proyecto de Don Juan José Landaburu sobre minería," Lima, 18 July 1827.

15. " . . . los mismos aviadores que corren los riesgos de esa especie de giro [crédito a la minería], lejos de hacer la fortaleza de aquellos [los mineros] causan su ruina." "Informe del Tribunal General de Minería sobre la minería peruana . . . ," AGN, SHMH, OL 233, ff. 593–637, 16 September 1834.

16. "Libro de Deslindes y Opposiciones," ADRMCP, lomo 81 (1840), see f. 33–36, records of the Tinyahuarco ingenio, June–August 1830.

17. See, for example, "Estancias y pueblos de indios en la region de Pasco," AGN, Serie Minería C-12, legajo 1786 (1786), which contains a sketch of the area that relates the "cerro de Colquijirca" to other localities and, finally, to the "laguna" of Chinchaycocha. See also Espinoza Claudio and Boza Monteverde, *Alcabalas y protesta,* for a discussion of the same document.

18. See Du Chatenet, *Estado actual,* 112–13.

19. "Corresponsal del 'Comercio,'" Cerro de Pasco, 10 January, in *El comercio* (Lima), Saturday, 15 January 1859, p. 3. The document acknowledged that the "bajas de 6 cajones de cascajo" cost 9 pesos per "cajón," a total of 54 pesos (downhill transportation of 6 boxes of ore).

20. "Informe del año 1897 sobre comercio y finanzas del Perú, por Alfred St. John," in Bonilla, ed., *Gran Bretaña y el Perú,* 1:282–83.

21. Of all the mints in nineteenth-century Peru, the one in Lima lasted longest, and throughout the century produced the largest flow of silver to the capital of the country. As for gold, the Cusco mint, at least between 1826 and 1839, was the largest consumer and transformer of gold metal into coins. See "Return of the Number of Marcs of Gold Coined at the Mints of Peru . . . ," PRO, FO (1834); and "Return of the Number of Marcs of Gold Coined in Peru in Each Year . . ." (1839), in *Accounts and Papers* of the *British Parliamentary Papers,* Statistics of Precious Metals, vol. 64 (1847), p. 207. See also Deustua, "Ciclo interno," 28–29, 38, and table 3.

22. See William M. Mathew, *Anglo-Peruvian Commercial and Financial Relations, 1820–1865.* Ph.D. dissertation, London: University of London, 1964, 77; Bonilla, ed. *Gran Bretaña y el Perú,* 5:96; Deustua, *Minería peruana,* 29–30.

23. F. C. Fuchs, "Mineral de Vinchos y oficina de Humanrauca," *Boletín de minas, industria, y construcciones* (Escuela Especial de Ingenieros de Lima, year 11, vol. 11, 1895).

24. Archive of Foreign Relations, Paris (Archives des Affaires Etrangères de Paris; hereafter AAEP), "Correspondance commerciale des consuls" (hereafter CCC); "Correspondance commerciale et consulaire"; "Communication du consul français à Islay," Islay, 23 May 1846.

25. Camprubí, *Historia de los bancos,* 356.

26. "Alumno Segundo Carrión," "Memoria del primer año," tesis no. 2, AUNI, Lima, 1878, p. 6 (1878); and "Alumno Pedro Félix Remy," "Memoria de Viage [*sic*]," tesis no. 1, Lima, 1878.

27. Germán Stiglich, *Diccionario geográfico del Perú*, 2d ed., 2 vols. (Lima: Imprenta Torres Aguirre, 1917–1922), 2:246.

28. On the setting up of bancos de rescate, see AGN, SHMH, PL 6, no. 114 (1826). It is argued there that 100,000 pesos in taxes and tributes would be necessary to establish a banco de rescate in Puno. See also AGN. SHMH. OL 186, caja 117, ff. 652–61. "Proyectos sobre bancos de rescate y reforma de las oficinas públicas presentados por D. Juan Evangelista Irigoyen" (1829); and OL 216 (1832). "La Diputación de Minería de Cerro de Pasco al Tribunal General de Minería," 16 February 1832, where it is said that "two hundred thousand pesos and 2,000 quintals of azogue were required every six months for the bank to operate in Pasco." In this same document it is also mentioned that the *Diputación de Minería* of Hualgayoc also demanded the establishment of a banco de rescate there.

29. See AGN, SHMH, PL 6, M 114 (1826); OL 186, caja 117, ff.652–61 (1825); OL 216 (1832).

30. See again ACNL, legajo 1 (1827); and "El Prefecto de este Departamento acompañando representación del Intendente de Chancay," AGN, SHMH, PL 6, no. 72 (1826).

31. On fears to contraband see "Informe del ensayador de las Cajas de Trujillo," AGN, SHMH, Prefectura de La Libertad, OL 197 (1830). See also OL 10, caja 2, f. 20 (1821); and "Informe de Francisco de Quirós, Prefecto de Junín, a los Diputados Territoriales de Minería del Cerro de Pasco," OL 224, Prefectura de Junín, caja 216, f. 1081, 20 March 1833. Finally, see the report of the British consul in Lima, Charles Milner Ricketts, in Bonilla, ed., *Gran Bretaña y el Perú*, 1:21, 29, which estimates the total illegal trade of mineral products at the beginning of the national period to have been between 1,229,000 and 1,232,000 pesos; and Deustua, *Minería peruana*, 45–54, which discusses the problem of contraband at the onset of the Peruvian republic.

32. "Informe de Nicolás de Piérola, Subdirector de Minería, al Ministro de Hacienda," AGN, SHMH, PL 6, no. 164, 2 August 1828.

33. Similar conflicts between the mineowners of neighboring towns, in the cases of Huallanca and Huari, could be seen in "La Diputación mineral de Huallanca pidiendo se dé por nula . . . ," AGN, SHMH, PL 10, no. 312 (1830).

34. Alberto Flores Galindo, *Aristocracia y plebe: Lima 1760–1830* (Lima: Mosca Azul Editores, 1984), 139–48. See also the essays by Carmen Vivanco Lara, Alberto Flores Galindo, Ward A. Stavig, Charles Walker, and Carlos Aguirre in Carlos Aguirre and Charles Walker, eds., *Bandoleros, abigeos, y*

montoneros: Criminalidad y violencia en el Perú, siglos dieciocho-veinte
(Lima: Instituto de Apoyo Agrario, 1990), 25–182.

35. Tschudi, *Testimonio del Perú*, 269. This was, of course, a heavily charged racist remark.

36. See "Proyecto de Don Juan José Landaburu . . . ," ACNL, legajo 1 (1827); "Informe del Tribunal General de Minería . . . ," AGN, SHMH, OL 233, ff. 593–637 (1834); see also "Proposición del minero Mariano Valderrama del asiento de San Antonio de Yauli . . . ," AGN, Serie Minería C-12, legajo 72 (1828); and "Informe del Prefecto de Junín, Francisco Quirós . . . ," AGN, SHMH, OL 224, f. 1069–75, 3 April 1833.

37. See Deustua, "Mining Markets, Peasants," esp. 38–39, 41–42.

38. Such was the case in the mining company of the Fernandini family as late as 1883 through 1889. See "Libro diario de la Negociación Minera del Dr. Erasmo Fernandini," Archive of the Agrarian Jurisdiction (Archivo del Fuero Agrario, Lima, hereafter AFA), Serie Algolán, ALG 195, 1883–1889, f. 119 and ff., where 176,367 soles were paid to llameros "por las bajas de mineral" (for downhill transportation of ores).

39. Manrique, "Arrieros de la sierra," 36.

40. AGN, Sección Casa de Moneda (hereafter SCM), República, legajo 92, CMR-00258b (1836).

41. Ibid., f. 6.

42. See AGN, SCM, República, legajo 92, CMR-00257, CMR-00258, CMR-00258a, and CMR-00258b (only for 1836). See also "Certificados por los ensayadores de la Casa de Moneda sobre recepción de plata para su amonedación," BN, Sala de Investigaciones Bibliográficas, D 901 (1826). See also AGN, SCM, República, legajos 101, 102, 103, and 104, CMR-00747, 00752–56, 00790–97, and 00830, for 1843.

43. AGN, SCM, República, legajo 92, CMR-00258 and CMR-00258a (1836).

44. Ibid.

45. Tschudi, *Testimonio del Perú*, 269.

46. Flores Galindo, *Aristocracia y plebe,* esp. 139ff. On the social and political significance of banditry and *montoneros* in the early republic, see also Charles Walker, "Montoneros, bandoleros, malhechores: Criminalidad y política en las primeras décadas republicanas," in Aguirre and Walker eds., *Bandoleros, abigeos, y montoneros,* 105–36.

47. AGN, SHMH, OL 197, ff. 1771–2175, Prefectura de la Libertad (1830).

48. AGN, SHMH, OL 163, caja 68, ff. 925–1155 (1827).

49. Another example of a large concentration of silver trading is that of Miguel Núñez in Puno. From the 15 bars smelted at the callana of Puno in March of 1831, 11 belonged to him, as did 9 from the 10 smelted the next month. Thus, in only two months he was dealing with 20 silver bars with a weight of 3,398 marcs and a value of 27,181 pesos. See AGN, SHMH, OL 207, ff. 1489–1622 (1831).

50. "Résolution du Congrès Général Constituant de la République du Pérou. Rapport du Consul, Lima, 26 avril 1828." AAEP, CCC, Lima, vol. 1, f. 241r.

51. Gootenberg, *Between Silver and Guano.*

52. Basadre, *Historia de la república,* 1:218.

53. See Chocano, *Comercio en Cerro de Pasco;* and "Circuitos mercantiles." Carlos Sempat Assadourian and a research team in Lima, including Carlos Contreras, Margarita Suarez, and Cristina Dam, were also working Custom Books, although they have not yet published their results. The information that these historical sources provide on the mining trade between 1819 and 1826 is very fragmented, and between 1826 and 1829 the alcabala records had only an informative character, not a tax one.

54. Chocano, *Comercio en Cerro de Pasco;* and "Circuitos mercantiles."

55. See also Deustua, "Routes and Roads," esp. 27–28.

56. See Chocano, *Comercio en Cerro de Pasco;* and "Circuitos mercantiles."

57. There are customs records in the AGN of 1829, for example.

58. On the impact of free trade on the Peruvian national economy in the nineteenth century and the presence of foreign merchants, see Bonilla, Del Rio, and Ortiz de Zevallos, "Comercio libre y crisis"; Magnus Morner, *Notas sobre el comercio y los comerciantes del Cusco desde fines de la colonia hasta 1930* (Lima: Instituto de Estudios Peruanos, 1979); and Gootenberg, "Social Origins" and *Between Silver and Guano.*

59. "Constancia del Conde de San Juan de Lurigancho," AGN, SCM, CMR-0034, Lima, 23 October 1821; and "Don Juan Begg, sobre que se afiance la cantidad de 60,000 pesos," AGN, SHMH, PL 6, no. 119 (1826).

60. Robert Proctor, "Cerro de Pasco y la explotación minera," in *El Perú visto por viajeros,* ed. Estuardo Núñez (Lima: Ediciones Peisa, 1973), 2:24–35, esp. 33. This article is only a small excerpt of his *Narrative of a Journey across the Cordillera of the Andes,* published originally in 1823.

61. An example of this competition, in the case of Cusco, can be seen in Bonilla, Del Rio, and Ortiz de Zevallos, "Comercio libre y crisis."

62. Another example of the conflict between Peruvian merchants and authorities and foreign traders is the "Hidalgo incident" of 1830, one of the outcomes of which was the blockade of the port of Callao by British vessels. The origin of the conflict was the trading of goods and money valued at 32,000 pesos, mostly mining goods such as silver bars, raw silver (plata piña), gold, copper, and coins, destined to John MacLean, a British merchant based in Lima. See Celia Wu Brading, *Generales y diplomáticos: Gran Bretaña y el Perú, 1820–1840* (Lima: Pontificia Universidad Católica del Perú, 1993), esp. 67–94.

63. "Rapport du Consul Général de France au Pérou, B. Barrère, au Ministre des Affaires Etrangères, à Lima, le 24 juillet 1830," AAEP, CCC, Lima, vol. 1, ff. 236r–v. For more on the internal conditions of domestic trade see the newspapers *Gaceta mercantil* and *Telégrafo de Lima*, both kept at BN, Lima, Sala de Investigaciones Bibliográficas.

64. See Gallagher and Robinson, "Imperialism of Free Trade"; William M. Mathew: "The Imperialism of Free Trade: Peru, 1820–1870," *Economic History Review* 21, 3 (London, 1968): 562–79; Bonilla, Del Rio, and Ortiz de Zevallos, "Comercio libre y crisis"; Gootenberg, "Social Origins," and *Between Silver and Guano;* Wu Brading, *Generales y diplomáticos.* On the role of the military in the early Republic, a key topic in Wu Brading's book, see also Víctor Villanueva, *Ejército peruano. Del caudillaje anárquico al militarismo reformista* (Lima: Librería-Editorial Juan Mejía Baca, 1973), esp. 40–42, on the confrontation with "English imperialism."

65. In AAEP, CCC, Lima, vol. 1, f. 236 and ff.

66. "Imprenta del Estado por J. Gonzalez," *El conciliador* (Lima), no. 54, Wednesday, 14 July 1830.

67. See AGN, SHMH, PL 6, nos. 8, 21, 30. These documents deal with the commercial operations of the merchant houses Cochran and Nixon Macall.

68. A great treatment of trade businesses in Lima, entrepôts, and ships controlled by merchants at the end of the eighteenth century is Flores Galindo, *Aristocracia y plebe,* esp. 54–84, and app. 3.

69. "Le gouvernement a donné des ordres pour qu'on y elevât des constructions, surtout des magazins, qu'il est dans l'intention de louer à un haut prix au commerce." AAEP, CCC, Lima, vol. 1, f. 238r.

70. Gootenberg, *Between Silver and Guano.*

71. Wu Brading, *Generales y diplomáticos,* also focuses on the role of the British consuls in the early Republic, particularly on Belford Hinton Wil-

son. Refuting Gootenberg's assertions in *Between Silver and Guano,* Wu Brading argues that the actions of British consuls were based on local circumstances and that they were not necessarily following the recommendations of the British Foreign Office.

72. "Correspondencia entre el Prefecto de Junín, Francisco Quirós, y el Señor Ministro de Hacienda," AGN, SHMH, OL 224, f. 1069 and ff. (1833). Quotations from the letter of 3 April 1833 by Francisco Quirós from Cerro (de Pasco). See also letters of 29 March and 13 April by "Señor Ministro de Hacienda," Martínez, from Lima.

73. On Quirós's involvement in the silver trade, see AGN SCM, República, legajo 92, CMR-00258b (1836); AGN, Protocolos Notariales, protocolo 965, Notario J. V. de Urbina, "Don Francisco Quirós," 15 July 1828, 43 v.; and protocolo 496, Notario Felipe Orellana, no. 28, f. 2966, 4, 30 November 1862 and 12 January 1863, "Juana Quirós and Francisco Quirós," on his personal and family wealth.

74. "Sumario que se le ha seguido al extranjero Don Patricio Ginez acusado de hacer por el puerto de Huacho el comercio clandestino de pastas de oro y plata." AGN, SHMH, PL 6, no. 318 (1826).

75. Ibid. ff. 1-7.

76. Ibid. ff. 12v-29.

77. Ginez's testimony and the last quotation are in ff. 16v-17.

78. The inspiration to ask my archival material such questions comes from reading the works of the great French historian Jean Meuvret, including *Etudes d'histoire économique* (Paris: Librairie Armand Colin, 1971), esp. 127-37; and Alphons Dopsch, *Economía natural y economía monetaria* (Mexico City: Fondo de Cultura Económica, 1943; original German edition, 1930). Meuvret's studies emphasize the nature of commercial exchanges in large trade business, where goods were exchanged against goods without the existence of money. They also demonstrate that only bad money, or account money, was present in small commercial exchanges, and that the extraction of surpluses in good money occurred in other commercial transactions. His work deals basically with the economic history of France in the sixteenth and seventeenth centuries.

79. For other situations in Latin America in which "vouchers, draft bills, or promissory notes" played a role as exchange mechanisms in the absence of money, see Teresita Martínez-Vergne, *Capitalism in Colonial Puerto Rico: Central San Vicente in the Late Nineteenth Century* (Gainesville: University Press of Florida, 1992), 80-85. Adam Százdi calls it "credit without

banking"; see Százdi, "Credit—without Banking—in Early-Nineteenth-Century Puerto Rico," *Americas* 19, 2 (October 1962): 149–71.

80. On the slow rotation of capital the classical reference is Karl Marx: "If the process of reproduction is slow, then so is the turnover of the merchant's capital." Marx, *Capital: A Critique of Political Economy,* 3 vols. (New York: International Publishers, 1967; original edition, Hamburg, 1894), 3:303. Concerning the Peruvian wool trade in the late nineteenth and early twentieth centuries, see Burga and Reátegui, *Lanas y capital mercantil,* 156–70.

81. On the geography of Peru, see Olivier Dollfus, *Le Pérou: Introduction géographique à l'étude du développement* (Paris: Institut des Hautes Etudes de l'Amérique Latine, 1968); see also Dollfus, "Les Andes intertropicales: Une mosaïque changeante," *Annales: Economies, sociétés, civilizations* n.s. 5–6 (Paris, September–December 1978): 895–905; and Dollfus, *El reto del espacio andino* (Lima: Instituto de Estudios Peruanos, 1981). On the Andean civilization, see Murra, *Formaciones económicas y políticas;* and Murra, *Organización económica.* On the Spanish conquest and its impact on the Andean pre-Columbian societies, see Wachtel, *Vision des vaincus;* and Romano, *Mécanismes de la conquête.*

82. On the raising and trading of mules see Nicolás Sánchez Albornoz, "La extracción de mulas de Jujuy al Perú: Fuentes, volumen, y negociantes," *Estudios de historia social,* no. 1 (Buenos Aires, 1965): 107–20; and Sánchez Albornoz, "La saca de mulas de Salta al Perú, 1778–1808," *Anuario del Instituto de Investigaciones Históricas,* no. 8 (Rosario, Argentina, 1965): 261–312. See also Assadourian, *Sistema de la economía,* esp. 40–53, 229–35.

83. In Concolorcorvo (Alonso Carrió de la Vandera), *Itinéraire de Buenos Aires à Lima,* trans. from the Spanish by Yvette Billod, with an introduction by Marcel Bataillon of the *Lazarillo de Ciegos Caminantes* (Paris: Institute des Hautes Etudes de l'Amérique Latine, 1961; original edition, Lima, 1776), 253.

84. "Memoria del Sub Prefecto de Jauja," in *El peruano* (Lima), 7 October 1874, cited in Manrique, "Arrieros de la sierra central," 33 (note 7).

85. Report of the British consul in Lima, Charles Milner Ricketts, in Bonilla, ed., *Gran Bretaña y el Perú,* 1:69.

86. Julio C. Avila and Ulises Bonilla, "Excursión a las minas de Parac y Colquipallana," AUNI, thesis no. 13 (17), report written 11 March 1889, studies 1884–1889.

87. J. M. McCulloch, *A Dictionary, Geographical, Statistical and Histor-*

ical of the Various Countries, Places, and Principal Natural Objects in the World, 2 vols. (London: Brown, Green and Longmans Publishers, 1846), 2:500. At the time this account was written, the American dollar and the Peruvian peso were approximately equal in value.

88. On the abundance of mules and horses in the valley of Tarma, see "Comunicación de Delgado, Intendencia de Tarma," AMRE, Prefecturas de Departamento (no precise date, probably September 1824), Z-O-E. Delgado gathered for the wars of Independence "some one hundred beasts, between mules and horses" in a few weeks.

89. Yauli, in the central sierras, was one of these *comunidades de pastores*, as was Yanacancha, in Cajamarca. See "Registro Cívico del Distrito de Yauli," ADRMH, 1883, ff. 13-24; José Antonio Araoz, "Excursión a Hualgayoc," AUNI, May 1889, thesis no. 25. For the twentieth century, see Jorge Flores Ochoa, *Los pastores de Paratía* (Mexico City: Instituto Indigenista Interamericano, 1968); Flores Ochoa, ed., *Pastores de puna [Uywamichiq punarunakuna]* (Lima: Instituto de Estudios Peruanos, 1977); Cipolletti, "Llamas y mulas."

90. Silvia Palomeque, "Loja en el mercado interno colonial," *Hisla*, no. 2 (Revista Latinoamericana de Historia Económica y Social, Lima, 1983): 33-45, esp. 37.

91. According to the "Arancel de repartimientos" of 1754, in Tord and Lazo, *Hacienda, comercio*, 152-58.

92. Ibid., 159.

93. Pflucker, *Exposición que presenta*, 9-12.

94. "Sumario que se le ha seguido . . . ," AGN, SHMH, PL 6, no. 318 (1826); Ricketts report in Bonilla, ed., *Gran Bretaña y el Perú*, 1:17-83.

Chapter 5

1. Alberto Regal, *Historia de los ferrocarriles de Lima* (Lima: Universidad Nacional de Ingeniería, 1965), 4-6.

2. Esteves, *Apuntes para la historia*, 143.

3. On Chorrillos and other places as fishing centers, see María Rostworowski de Diez Canseco, *Recursos naturales renovables y pesca, siglos dieciseis y diecisiete* (Lima: Instituto de Estudios Peruanos, 1981); Alberto Flores Galindo, "La pesca y los pescadores en la costa central (siglo dieciocho)," *Histórica* 5, 2 (Lima, December 1981): 159-65; and Flores Galindo,

Aristocracia y plebe, 186–96. On Chorrillos, Barranco, and Miraflores as beach resorts, see José Antonio Del Busto Duthurburu, *Historia y leyenda del viejo Barranco* (Lima: Editorial Lumen, 1985); and Jorge Avendaño Hübner, *Miraflores de antaño* (Lima: Universidad Peruana Cayetano Heredia, 1989). On the spiritual discovery of the sea, see Alain Corbin, *The Lure of the Sea: The Discovery of the Seaside in the Western World, 1750–1840,* trans. from the French by Jocelyn Phelps (Berkeley: University of California Press, 1994).

4. J. B. H. Martinet, *La agricultura en el Perú* (Lima: Centro Peruano de Historia Económica, Universidad Nacional Mayor de San Marcos, 1977; original edition, 1877), 97; Basadre, *Historia de la república,* 3:180–2, 322, and 4:313; Regal, *Historia de los ferrocarriles.*

5. Jacinto López, *Manuel Pardo* (Lima: Editorial Gil, 1947), 299–300.

6. Yepes del Castillo, *Perú,* 97–101; Bonilla, *Guano y burguesía,* 53–60.

7. Martinet also presented this report, under the title *L'agriculture au Pérou,* to the Congrès Internationale de l'Agriculture that took place in Paris in 1878. The original *Mémoire* is at the National Library in Paris (Bibliothèque Nationale de Paris). Although I have consulted this original copy, I have worked instead with the more recent Peruvian edition, J. B. H. Martinet: *La Agricultura en el Perú,* Lima: Centro Peruano de Historia Económica, Universidad Nacional Mayor de San Marcos, published by Pablo Macera and Honorio Pinto in 1977.

8. In order to establish a comparison, let me mention that in France in 1842 there were already 500 kilometers of railroads, which increased to 2,000 kilometers in 1848. André Lefevre adds: "A la fin de 1852, la longueur des lignes exploitées est de 3,886 kilomètres et celle des lignes concédées de 6,912." André Lefevre, *Sous le second empire: Chemins de fer et politique* (Paris: Société d'Edition d'Enseignement Supérieur, 1951), 13. For a general comment on French railroads, see Pierre Goubert, *Initiation à l'histoire de la France* (Paris: Fayard and Tallandier, 1984), 322. In the United States in 1840 2,818 miles of railroads had already been built. This increased to 9,021 miles in 1850 and to 30,636 in 1860, see Slason Thompson, *A Short History of American Railways* (New York: D. Appleton and Company, 1925), 97. For a general comment on railroads in the United States, see John Garraty, *A Short History of the American Nation* (New York: Harper and Row, 1981), 207. Peru was clearly well behind France and the United States in railroad construction.

9. See Lautaro Núñez, "L'évolution millénaire d'une vallée: Peuplement

et ressources à Tarapacá," *Annales: Economies, sociétés, civilizations,* n.s. 5-6 (Paris, September-December 1978): 906-20, esp. 918. See also Antonio Raimondi, *El Perú,* 2d ed. (Lima: Librería e Imprenta Gil, 1902), 4:514; and Leon E. Bieber, "Empresarios mineros en el siglo diecinueve: Bases para su caracterización social," Instituto de Estudios Sociales y Económicos, Cochabamba, Bolivia, 1980 (mimeo), which deals with the neighboring province of Atacama.

10. Dirección de Estadística, *Estadística de las minas,* 92-93; and Emilio Dancuart and J. M. Rodríguez, "Estado de la industria," in *Anales de la hacienda pública del Perú (1821-1889),* 19 vols. (Lima: Imprenta de La Revista, 1902-1926), 17:95-96.

11. See José Matos Mar and Fernando Fuenzalida, "Proceso de la sociedad rural," in *Hacienda, comunidad, y campesinado en el Perú,* ed. José Matos Mar (Lima: Instituto de Estudios Peruanos, 1976), 15-50, esp. 31-32.

12. See Margarita Giesecke, *Masas urbanas y rebelión en la historia: Golpe de Estado, Lima, 1872* (Lima: Centro de Estudios y Divulgación de Historia Popular, 1978), 75, 79, 81.

13. J. B. H. Martinet, *Carestía de víveres en Lima* (Lima: Centro Peruano de Historia Económica, Universidad Nacional Mayor de San Marcos, 1977; original edition, 1875).

14. Martinet, *Agricultura en el Perú,* 95. The population of Lima increased from 52,627 in 1796 to 94,195 in 1857, 120,994 toward 1876, and 130,089 in 1903. Apart from the references in Giesecke, *Masas urbanas,* see Henry Dobyns and Paul L. Doughty, *Peru, a Cultural History* (New York: Oxford University Press, 1976), 298-99 (table 1). Their data, however, differ from that of Giesecke and from Centro de Estudios de Población y Desarrollo, *Informe demográfico del Perú* (Lima: CEPD, 1972), 92.

15. Bonilla, *Guano y burguesía,* 61-108; Tantaleán Arbulú, *Política económico-financiera,* 68-115.

16. Watt Stewart, *Henry Meiggs, Yankee Pizarro* (Durham, N.C.: Duke University Press, 1946).

17. Spenser St. John (British consul in Peru), in Bonilla, ed., *Gran Bretaña y el Perú,* 1:197.

18. Martinet, *Agricultura en el Perú,* 94-98; Esteves, *Apuntes para la historia,* 141-46; Basadre, *Historia de la república,* 6:166-72.

19. Piel offers the following data: 24 kilometers of railroads were built in 1851, 87 km. in 1856, 103 km. in 1857, 138 km. in 1868, 255 km. in 1869, 668 km. in 1870, 1,792 km. in 1875, 2,030 km. in 1877. The amount of new track

dropped to 1,509 km. in 1883, but slowly recovered to 1,580 km. in 1884, 1,598 km. in 1890, 1,734 km. in 1895, and 1,799 km. in 1900. See Jean Piel, *Capitalisme agraire au Pérou: L'Essor du neo-latifundisme dans le Pérou republicain* (Paris: Editions Anthropos, 1983), 63. See also the very accurate and descriptive book, almost a testimony, by Federico Costa y Laurent, *Reseña histórica de los ferrocarriles del Perú* (Lima: Litografía y Tipografía Carlos Fabri, 1908).

20. Flores Galindo, *Arequipa y el sur andino*, esp. 82–89.

21. Rory Miller, "Railways and economic development in central Peru, 1890–1930," in *Social and Economic Change in Modern Peru*, ed. Rory Miller, Clifford T. Smith, and John Fisher (Liverpool: University of Liverpool, 1976), 29–30. See also Yepes del Castillo, *Perú*, 147.

22. Yepes del Castillo, *Perú*, 137–40; Cotler, *Clases, estado, y nación*, 125–26; Mallon, *Defense of Community*, 129; Clayton, *Grace*, 141–75; Miller, "Grace Contract," 73–100.

23. In 1927 the Southern Railway "linked the city of Cusco with Machu Picchu [110 km.], and by 1951 had reached Huadquiña, in the La Convención valley [131 km.]. Today the railroad extends to Chaullay, 25 kilometers beyond Quillabamba." Eduardo Fioravanti, *Latifundismo y sindicalismo agrario en el Perú* (Lima: Instituto de Estudios Peruanos, 1974), 22.

24. Dollfus, *Pérou*, 249.

25. On the presence of foreign capital in the mining economy in early-twentieth-century Peru, see DeWind, "Peasants Become Miners," chap. 1; Mallon, *Defense of Community*, 170–74; Dirk Kruijt and Menno Velinga, *Estado, clase obrera, y empresa transnacional: El caso de la minería peruana, 1900–1980* (Mexico City: Siglo Veintiuno Editores, 1983); and Dore, *Accumulation and Crisis*. On Empresa Socavonera del Cerro de Pasco, see Yepes del Castillo, *Perú*, 149.

26. See, for example, Fuchs, "Mineral de Vinchos"; Mallon, *Defense of Community*, 170ff.; Caballero Martín, *Imperialismo y campesinado en la sierra central* (Huancayo, Peru: Instituto de Estudios Andinos, 1981), 15–18, 72–96.

27. See Stewart, *Henry Meiggs*, the third chapter of which bears the title "Railroad Fever."

28. On the structure and agrarian evolution of the valleys of the Peruvian northern coast in the nineteenth century, see Klarén, *Modernization, Dislocation, and Aprismo* (which describes the Chicama valley and the city of Trujillo); Burga, *Encomienda a la hacienda capitalista* (on the Jequete-

peque valley and the port of Pacasmayo); and Gonzales, *Plantation Agriculture and Social Control.*

29. Mallon, *Defense of Community;* Manrique, *Mercado interno y región.* See also Gerardo Rénique, "El desarrollo de una empresa ganadera en los Andes centrales (1910–1960)," *Tierra y Sociedad,* 1.1 (Revista del Archivo del Fuero Agrario, Lima, April 1978), 39–59.

30. Gonzales, *Plantation Agriculture and Social Control,* 1–69, esp. 12.

31. See Macera, "Plantaciones azucareras," 4:231–34.

32. Dollfus, *Pérou,* 249.

33. Miller, "Railways and Economic Development," 35–36. The best work on the Mexican railroads and their impact on the Porfirian economy is John H. Coatsworth, *Growth against Development: The Economic Impact of Railroads in Porfirian Mexico* (De Kalb: Northern Illinois University Press, 1981). Table 2.3 in that book shows "the expansion of Railroads under Federal Concession" between 1873 and 1910. At that time Mexican rail lines increased from 572 to 19,205 kilometers (36–37).

34. See, for example, Valderrama and Escalante, "Arrieros, troperos y llameros"; and Cipolletti, "Llamas y mulas." See also Benjamin Orlove, "Alpaca, Sheep, and Men: The Wool Export Economy and Regional Society in Southern Peru" (Ph.D. dissertation, University of California, Berkeley, 1974).

35. See Antonio Raimondi, "Memoria sobre el Cerro de Pasco y la montaña de Chanchamayo," originally published in 1885, later reprinted in his collection *El Perú* (Lima: Editores Técnicos Asociados, 1965), 4:444–88; Fiona Wilson, "The Dynamics of Change in an Andean Region: The Province of Tarma, Peru, in the Nineteenth Century" (Ph.D. dissertation, University of Liverpool, 1978); Fiona Wilson: "Propiedad e ideología: Estudio de una oligarquía en los Andes centrales (siglo diecinueve)," *Análisis* 8–9 (Cuadernos de Investigación, Lima, 1979): 36–54, esp. 48; Manuel Manrique: "La colonización y la lucha por la tierra en el valle del Perené," in *Los movimientos campesinos en el Perú, 1879–1965,* ed. Wilfredo Kapsoli (Lima: Delva Editores, 1977), 267–300; and Mallon, *Defense of Community,* 49, 59, 60, 126, 128, 137–38, 165.

36. See, for example, Carlos Iván Degregori, *Ayacucho, raíces de una crisis* (Ayacucho: Instituto de Estudios Regionales José María Arguedas, 1986); and Degregori, *Ayacucho 1969–1979: El surgimiento de Sendero Luminoso* (Lima: Instituto de Estudios Peruanos, 1990).

37. Gordon Appleby, "Las transformaciones del sistema de mercados en

Puno: 1890–1960," *Análisis* 8–9 (Cuadernos de Investigación, Lima, May–December 1979): 55–71, 57. See also Appleby, "Exportation and Its Aftermath: The Spatial Economic Evolution of the Regional Marketing System in Highland Puno, Peru" (Ph.D. dissertation, Stanford University, 1978).

38. On the export crops of the Peruvian coast, see Klarén, *Modernization, Dislocation, and Aprismo;* Burga, *Encomienda a la hacienda capitalista;* Gonzales, *Plantation Agriculture and Social Control;* and Macera, "Plantaciones azucareras." On the integration of the central sierra to the economic dynamic of Lima, see Mallon, "Minería y agricultura" and *Defense of Community;* Manrique, *Mercado interno y región.* On the Andean south, see Heraclio Bonilla, "Islay y la economía del sur peruano en el siglo diecinueve," in *Mecanismos de control,* 105–21; Flores Galindo, *Arequipa y el sur;* Burga and Reátegui, *Lanas y capital mercantil;* and Jacobsen, *Mirages of Transition.*

39. In addition to the works already mentioned on the history of railroad building, see Costa y Laurent, *Reseña histórica,* a well-balanced study of the state of railroad building at the beginning of the twentieth century.

40. Martinet, *Agricultura en el Perú,* 94.

41. Dirección de Estadística, *Estadística de las minas,* 92–93; and Ministerio de Hacienda, *Padrón general de minas.*

42. See *Ministro de Gobierno, Policía, y Obras Públicas: Memoria que presenta al Congreso Ordinario de 1874* (Lima: Imprenta de El Comercio, 1874), 31.

43. Carlos Contreras, "Mineros, arrieros, y ferrocarril en Cerro de Pasco, 1870–1904," *Hisla,* no. 4 (Revista Latinoamericana de Historia Económica y Social, Lima, 1984): 3–20, esp. 8–9.

44. Ismael C. Bueno, "Asiento del Cerro de Pasco," *Boletín de minas, industria, y construcciones* 3, 3 (Lima: Escuela Especial de Ingenieros de Minas, 1887).

45. Francisco R. del Castillo, "Excursión a Huarochirí y Yauli," AUNI, thesis no. 40 (1892).

46. Miller, "Railways and Economic Development."

47. Contreras, "Mineros, arrieros, y ferrocarril," esp. 3.

48. Similar arguments have been made comparing the domestic building of the *socavón de desague* of Quiulacocha and the importing of steam engines by the British Steam Engine Company between 1820 and 1840 in Cerro de Pasco. See Deustua, "The Socavon de Quiulacocha and the Steam Engine Company: Technology and Capital Investment in Cerro de Pasco, 1820–

1840," in Miller, ed., *Region and Class*, 35–75, esp. 63–67. There I state that it is necessary to discuss "a more general theoretical (and also practical) problem: that of the social and economic conditions in which certain technologies, originating in other economic and social milieux, can be applied and, above all, function for the balanced and overall development of our nations (dixit Peru)," 67.

49. Miller, "Railways and Economic Development."

50. Ministro de Gobierno, *Memoria que presenta al Congreso*, 31.

51. Basadre, *Historia de la república*, 6:108–9.

52. "Excursión a Parac" by Carlos Y. Lissón, AUNI, thesis no. 47 (1893).

53. For ideological debates on Peruvian economic development, see Gootenberg, *Imagining Development*.

54. Michel Fort, "Informe sobre la mina San Antonio de Bellavista-Huarochirí," AUNI, thesis, January to March 1889 (Studies 1884–1890) (1890).

55. Tschudi, *Testimonio del Perú*, 233.

56. Baldomero Aspíllaga, "Excursión a Huarochirí," AUNI, thesis no. 31 (1889).

57. "Libro diario de la Negociación Minera del Dr. Erasmo Fernandini," AFA, Lima, Serie Algolán, ALG 195, years 1883–1889. On the importance of Fernandini's mining and agriculture enterprises for the region, see Juan Sánchez Barba, "La via terrateniente y campesina en el desarrollo capitalista en la sierra central: El caso de Cerro de Pasco," in *Campesinado y capitalismo*, ed. Juvenal Casaverde et al. (Huancayo: Instituto de Estudios Andinos, 1979), 147–234; Deustua, "Mines et monnaie," 2:373–530; and Mallon, *Defense of Community*, 136–37, 172–73.

58. José Antonio Araoz, "Excursión a Hualgayoc," AUNI, thesis no. 25, May 1889 (studies 1886–90; graduation 1890).

59. Ibid. See also F. Málaga Santolalla, *Recursos minerales de la provincia de Hualgayoc*, boletín no. 6 (Lima: Cuerpo de Ingenieros de Minas, 1902).

60. See also Taylor, *Main Trends;* Taylor, "Earning a Living"; and Málaga Santolalla, *Recursos minerales*.

61. José Antonio Araoz, "Excursión a Hualgayoc," AUNI, thesis no. 25, May 1889 (studies 1886–90; graduation 1890).

62. See Mallon, *Defense of Community*. See also Deustua, "Mining Markets, Peasants, and Power"; and, on the issue of peasant resistance in the Andes (Bolivia in this case), see Tristan Platt, "The Andean Experience of

Bolivian Liberalism, 1825–1900: Roots of Rebellion in Nineteenth-Century Chayanta (Potosí)," in Stern, ed., *Resistance, Rebellion,* 280–323.

63. Miller, "Railways and Economic Development," 42.

64. Ismael C. Bueno, "Informe sobre el asiento mineral del Cerro de Pasco," in *Boletín de Minas, Industria y Construcciones.* Lima: Escuela Especial de Ingenieros, year 8, vol. 8 (1892).

65. Ibid.

66. Ibid.

67. Francisco R. Del Castillo, "Excursión a Huarochirí y Yauli," AUNI, thesis no. 40 (studies 1891, thesis 1892).

68. Ibid.

69. Ibid.

70. Julio C. Avila and Ulises Bonilla, "Excursión a las minas de Parac y Colquipallana," AUNI, thesis no. 13 (17), report written 11 March 1889 (studies: 1884–89).

71. Julio A. Morales, "Excursión a Huarochirí," AUNI, thesis no. 42, report written 11 June 1892 (1892).

72. See again "Libro diario de la Negociación . . . ," AFA, Serie Algolán, ALG 195.

73. Julio A. Morales, "Excursión a Huarochirí," AUNI, thesis no. 42, report written 11 June 1892 (1892).

74. In a previous work I wrote: "Nevertheless it is true, as the references show, that there was complementarity between the economic processes of mines and of livestock *estancias,* and not necessarily between those of mines and of agricultural haciendas, particularly because the former were located in the highest altitudes of the puna, as was the case of Cerro de Pasco, Morococha and, in part, Casapalca. Furthermore, this is true because the livestock of the estancias were mostly cameloids, an indispensable means of transportation for Andean mining, although not as good as the mules that came from the northwest of Argentina." José Deustua, "La minería, las clases sociales, y la independencia del Perú," *Análisis* 12 (Cuadernos de Investigación, Lima, 1983): 50–62, esp. 58. On the herding of cameloids on the puna, see among others, Flores Ochoa, ed., *Pastores de puna.*

75. Julio A. Morales, "Excursión a Huarochirí," AUNI, thesis no. 42, report written 11 June 1892 (1892).

76. Report by the British consul, Alfred St. John, in Bonilla, ed., *Gran Bretaña y el Perú,* 1:282–83.

77. Bueno, "Informe sobre el asiento mineral," 1892.

78. Araoz, "Excursión a Hualgayoc," 1889.
79. Nine to ten days, according to Contreras, "Mineros, arrieros, y ferro-carril," 6.
80. Miller, "Railways and Economic Development."

Chapter 6

1. See Gramsci, *Prison Notebooks;* for the historical use of some Gramscian ideas in nineteenth-century Mexico and Peru, see Mallon, *Peasant and Nation.*
2. See, for example, Gootenberg, *Imagining Development,* esp. chaps. 2, 4, and 5, on the discussion and alternatives for Peruvian economic development in the nineteenth century: from "heterodox traditions" and "popular industrialism" to the predominant "export liberalism."
3. Between 1873 and 1879, for example, the German government put 7,104,895 ounces of silver from its monetary reserves into the market, contributing to the collapse of silver prices and the international movement to the gold standard. See Mitre, *Patriarcas de la plata,* 195, app. 3; see also Deustua, "Minería a la acuñación," esp. 79–80, 119–20. See additional references in chapter 2, above.
4. Umberto Eco, with Richard Rorty, Jonathan Culler, and Christine Brooke-Rose, *Interpretation and Overinterpretation,* ed. Stefan Collini (Cambridge: Cambridge University Press, 1992), 25. Again, some recent examples of postmodernist analysis in Latin American history are: Joseph and Nugent, eds., *Everyday Forms of State Formation;* Mallon, *Peasant and Nation;* and Becker, *Setting the Virgin on Fire.*
5. Umberto Eco, *The Open Work,* trans. from the Italian by Anna Cancogni, with an introduction by David Robey (Cambridge, Mass.: Harvard University Press, 1989).
6. On preindustrial or protoindustrial economic cycles, see Pierre Deyon, "L'enjeu des discussions autour du concept de 'protoindustrialisation,'" *Revue du nord,* 61, 240 (1979): 9–18; and Mendels, "Industries rurales." On Latin American pre- and protoindustrialization, particularly concerning *obrajes,* see Manuel Miño Grijalva, "Capital comercial y trabajo textil: Tendencias generales de la protoindustria colonial latinoamericana," *Hisla,* no. 9 (Lima, 1987): 59–79; Richard J. Salvucci, *Textiles and Capitalism in Mexico: An Economic History of the Obrajes, 1539–1840* (Princeton:

Princeton University Press, 1987); and Miriam Salas de Coloma, *De los obrajes de Canaria y Chincheros a las comunidades indígenas de Vilcashuamán, siglo dieciseis* (Lima: n.p., 1979).

7. On the role of foreign capital in mining at the end of the nineteenth and at the beginning of the twentieth centuries, see Flores Galindo, *Mineros de la Cerro de Pasco,* 21–34; Mallon, *Defense of Community,* 169–74; and Caballero Martín, *Imperialismo y campesinado,* 15–17 and 72–106; DeWind, "Peasants Become Miners"; Dore, *Accumulation and Crisis;* and Kruijt and Vellinga, *Estado, clase obrera, y empresa transnacional.*

8. Ricardo Bentín was already working mines in Huarochirí in the 1870s. His association with John Howard Johnston of Bath (New Hampshire) and Jacob Backus of Brooklyn could have started then. See Deustua, "Mines, monnaie," chapter 5, esp. 532–40; and Sánchez, *Historia de una industria.*

9. See, among others, Klarén, *Modernization, Dislocation, and Aprismo;* Burga, *Encomienda a la hacienda,* esp. chap. 6, 163–201; Gonzales, *Plantation Agriculture and Social Control;* and Macera, "Plantaciones azucareras."

10. Thorp and Bertram, *Peru 1890–1977;* Mallon, *Defense of Community,* esp. chaps. 4, 5, and 6.

11. See, Hunt, *Growth and Guano;* Boloña, "Tariff Policies" and "Perú"; Macera, "Plantaciones azucareras," 99–100 (table 6), 150–228; Gootenberg, *Between Silver and Guano,* app. 2, table 2.1, and "Merchants, Foreigners, and the State"; finally, Quiroz, *Domestic and Foreign Finance,* esp. app. A, table A.1, 219–21.

12. See note 9, above. Finally, on guano the three best references are: Levin, *Export Economies;* Hunt, *Growth and Guano;* and Bonilla, *Guano y burguesía.*

13. See Camprubí, *Historia de los bancos;* and Quiroz, *Domestic and Foreign Finance.*

14. On the *bancos de rescate* see also Camprubí, *Bancos de rescate;* and Deustua, "Minería a la acuñación," esp. 87–93.

15. See Deustua, "Socavón of Quiulacocha."

16. See Mathew, *House of Gibbs;* Clayton, *Grace;* C. Alexander G. De Secada, "Armas, guano, y comercio marítimo: Los intereses de W. R. Grace en el Perú, 1865–1885," *Hisla,* no. 7 (Revista Latinoamericana de Historia Económica y Social, Lima, 1986): 105–29; Stewart, *Henry Meiggs;* and Bonilla, *Guano y burguesía,* chap. 2, 61–108. W. R. Grace was an Irish immigrant who came to Peru in 1850 when he was eighteen years old. He worked first as an employee of John Bryce, a merchant based in Callao. In

1860 he was already a partner in Bryce, Grace and Co., and by 1862 he had acquired a small fortune ($180,000). He settled later in New York and became mayor of the city, played a key role in the negotiations of the Grace Contract to eventually build the first multinational corporation in Latin America, which included interests in sugar, textiles, mining, paper and chemical products, and rail, sea, and air transportation. De Secada suggests that the role of the Grace Company in Peru could be compared to that of the United Fruit company in Central America.

17. Hunt, in *Growth and Guano,* asserts that railroads were "a risky overinvestment." See also Gootenberg, *Imagining Development,* 89 and 108, who qualifies Hunt's comment as modest.

18. Basadre, *Historia de la república,* vols. 3 and 4; Gootenberg, *Imagining Development.*

19. See, Levin, *Export Economies;* Hunt, *Growth and Guano;* and Bonilla, *Guano y burguesía.* Specifically for a discussion of production and productivity, see Hunt, *Growth and Guano,* 84; and Bonilla, *Guano y burguesía,* 137–38.

20. Deustua, "Producción minera y circulación monetaria," 338, emphasis added.

21. On the Grace Contract see again Yepes del Castillo, *Perú,* 137–40; Cotler, *Clases, estado, y nación,* 125–26; Mallon, *Defense of Community,* 129; Clayton, *Grace,* 141–75; and Miller, "Grace Contract." The estimation of the Peruvian foreign debt at £51 million comes from Guillermo Billinghurst, *Mensaje al congreso, 1913,* quoted in Yepes del Castillo, *Perú,* 138.

22. See Deustua, "Socavón of Quiulacocha."

23. As the reader may notice, the argument is now fully Weberian. See Max Weber, *The Protestant Ethic and the Spirit of Capitalism.* trans. from the German by Talcott Parsons (New York: Charles Scribner's Sons, 1958; original edition, Tübingen: J. C. B. Mohr, 1904–5).

24. See, for example, Richard M. Morse, "The Heritage of Latin America," in *The Founding of New Societies,* ed. Louis Hartz (New York: Mentor Books, 1964). See also Morse, *El espejo de Próspero* (Mexico City: Siglo Veintiuno Editores, 1982). A recent study of the 1988 Mexican presidential election argues that "the traditional Mexican political system" combines "a hierarchical system with an individual one." The dominant hierarchical system, of course, has its roots in colonial times. See Larissa Adler Lomnitz, Claudio Lomnitz Adler, and Ilya Adler, "The Function of the Form: Power Play and Ritual in the 1988 Mexican Presidential Campaign," in

Constructing Culture and Power in Latin America, ed. Daniel H. Levine (Ann Arbor: University of Michigan Press, 1993), 357–401, esp. 368–69.

25. Anthony B. Anderson, Peter H. May, and Michael J. Balick, *The Subsidy from Nature: Palm Forests, Peasantry, and Development on an Amazon Frontier* (New York: Columbia University Press, 1991). The authors raise the question of how to calculate the costs of depleting the natural environment, of taking for free "natural goods" (minerals in my case study, trees in theirs) that are part of a physical and social environment. If Anderson, May, and Balick are concerned about the deforestation of babassu trees in the Amazon area of Bolivia and Brazil, particularly in the state of Maranhao, my book poses the same question for a process of demineralization in nineteenth-century Peru.

26. On Raimondi and mining in nineteenth-century Peru, see Deustua, *"Routes, Roads,"* 21–22.

Glossary

acomodana	payment of mine workers in goods and food-stuffs
alcabala	sales tax
al fiado	sales on credit
apire	type of mine worker, porter
arriería	muleteering
azogue	mercury
banco de rescate	government bank that purchases silver metals
barretero	type of mine worker, digger, works with an iron bar
bocamina	a superficial mine, not a deep one
callana	government smelting plant
caporal	foreman
carrero	type of mine worker, pulls the wagons or carts
casa de fundición	same as *callana*
cascajo	gravel, mineral nuggets
chacanía	working day of a llama driver or *llamero*
circo	mining yard in which silver ores are combined with mercury, usually having a circular shape
criandero	animal raiser, whether of llamas, sheep, or mules
estancia	large livestock farm

enganche	system of recruiting forced or induced labor
ferrocarril	railroad
hacienda	large agricultural estate
huarache	a working night shift between day shifts in which the mine worker is given and chews coca leaves to fight sleep and fatigue
ingenio	silver refining mill
jornalero	worker, usually a wage laborer working for the day
llamero	llama driver
maquipurero	temporary worker
matricula	register
mayordomo	overseer
minero	miner
minería	mining
oficina	plant
operario	worker
padrón	census
parada	stop site
patio	yard, applies to the yard or the system which refines silver with the use of mercury
peón	peon, worker under strict control
plata piña	silver metal obtained after the processing of silver ores with mercury in *patios* and *circos*, it has the shape of a pineapple
puna	upper levels of the Andean mountains
real	a fraction of the peso, 8 reales equals one peso

quintal	a measure of weight, equal to 100 pounds
socavón	deep mine with multiple tunnels and shafts
sol	new Peruvian currency established in 1863; one sol was equal to one peso, but it was divided into 100 cents or 100 *centavos*
vara	a measure of length equal to 0.836 meters

Bibliography

Primary Sources

Archives

Archivo General de la Nación, Lima (AGN)

—Libros Manuscritos Republicanos H-4

Documents: 0450 (1849).

—Sección Histórica del Ministerio de Hacienda (SHMH)

Documentos Particulares. PL 6, nos. 8, 21, 30, 72, 114, 119, 164, 177, 193–194, 244, 318, 331 (1826); PL 7, nos. 46, 225 (1827); PL 10, no. 312 (1830).

Documentos Oficiales. OL 10, caja 2 (1821); OL 163, caja 68, (1827); OL 186, caja 117 (1829); OL 207 (1831); OL 216 (1832); OL 225 (1833); OL 233 (1834).

Tesorería Departamental. OL 40, caja 6 (1821).

Prefectura de La Libertad. OL 131, caja 38 (1825); OL 197 (1830).

Prefectura de Junín. OL 224, caja 216 (1833).

Dirección General de Minería. OL 164, OL 175, OL 186, caja 117 (1826–28).

Tribunal Mayor de Cuentas. OL 8, caja 1 (1821).

—Sección Casa de Moneda, República

Legajo 74. Expedientes CMR-0034 (1821); Legajo 83. Expedientes CMR 0095, CMR 0098, CMR 0099, CMR 00100, CMR 00101 (1828–29).

Legajo 92. Expedientes CMR-00257, CMR-00258, CMR-00258a, CMR-00258b (1836).

Legajos 101, 102, 103, 104. Expedientes CMR 00747, 00752–56, 00790–97, 00830 (1843).

—Sección Protocolos Notariales

Escribano Francisco Palacios. Protocolo 591 (1879).

Escribano José de Selaya. Protocolo 700 (1838). Testamento de José Lago y Lemus.

Escribano Manuel Suárez. Protocolos 876 (1807), 880 (1820). Poderes de Pedro Abadía.

Notario J. V. de Urbina. Protocolo 965 (1828).

Notario Felipe Orellana. Protocolo 496 (1862–63).

—Serie Impresos H-G

Boletines del Cuerpo de Ingenieros de Minas del Perú. H-G-0621 to H-G-0628.

Padrones Generales de Minas, 1899–1920. H-G-0718 to H-G-0729.

Registro Oficial de Fomento, Minas e Industria, 1901–6. H-G-0752 to H-G-0776.

Extracto Estadístico del Perú, 1920–30. H-G-0410 to H-G-0413.

—Serie Minería C-12

Legajos 1786 (1786); 61 (1825–28); 71 (1828); 72 (1828–47); 74 (1833); 76 (1846).

—Serie Tributos

Legajo 6, cuaderno 179 (1827).

Biblioteca Nacional del Perú, Lima (BN)

—Sala de Investigaciones Bibliográficas

—Serie Manuscritos Republicanos

Documents. D 10363 (1825); D 8696 (1826); D 6772 (1826); D 901 (1826); D 9465 (1827); D 9510 (1840); D 1696 (1849); E 1148 (1907); E 479 (1917); E 929 (1925); E 992 (1929).

Archivo de la Universidad Nacional de Ingeniería, Lima (AUNI)

—Memorias de Viage. Pedro Félix Remy, Cajavilca, Ica (1878); Segundo Carrión, Otuzco, Ica, Salpo, and Huamantanga (1878–79); Juan Garnier, Salpo, Otuzco, Ica, and Canta (1879); Ismael Bueno, Yauli (1885); Federico Villareal, Yauli (1885); Ismael Bueno, Cerro de Pasco (1887);

Germán Remy, Cerro de Pasco (1887); José Antonio Araoz, Hualgayoc (1889); Baldomero Aspíllaga, Huarochirí (1889); Julio C. Avila and Ulises Bonilla, Parac and Colquipallana (1889); Celso Herrera and Felipe A. Coz, Huarochirí (1889); Michel Fort, Huarochirí (1890); Francisco R. del Castillo, Huarochirí and Yauli (1891–92); Julio A. Morales, Huarochirí (1892); Carlos Y. Lisson, Parac (1893); Santiago Marrau, Huarochirí (1894).

—Sección Tesis Estudiantiles y Manuscritos

—Tesis 1 (1878); 2 (1878); 13 (1884); 14 (1885); 17 (1889); 25 (1890); 31 (1889); 40 (1892); 42 (1892); 47 (1893); 48 (1894).

Archivo del Fuero Agrario, Lima (AFA)

—Serie Algolán. ALG 205, Correspondencia de Manuel Clotet con Eulogio Fernandini (1898); ALG 195, Libro Diario de la Negociación Minera del Dr. Erasmo Fernandini (1883–89).

Archivo de la Dirección Regional de Minería de Huancayo, Huancayo, Perú (ADRMH)

—Registro Cívico del Distrito de Yauli, 1883.

Archivo de la Dirección Regional de Minería del Cerro de Pasco, Cerro de Pasco, Perú (ADRMCP)

—Libro Copiador de Notas desde 1832 hasta 1835. Correspondencia.

—Libro de Deslindes y Oposiciones (1840).

Archivo Legal de la Empresa Minera Centro-Min Perú, Lima (ALCMP)

—Documents. Legajo n. 45 (1896).

Archivo del Museo Nacional de Historia, Lima (AMNH)

—Manuscrito 2082 (1844).

Archivo del Ministerio de Relaciones Exteriores, Lima (AMRE)

—Prefecturas de Departamento (1824), Z-O-E.

Archivo del Congreso Nacional, Lima (ACNL)

—Legajo 1, n. 16. "Proyecto de Don Juan José Landaburu sobre minería" (1827).

Public Record Office (PRO), Foreign Office (FO), London

—Accounts and Papers of the British Parliamentary Papers. Vol. 64 (1847); vol. 39 (1849); vol. 76 (1896).

—Consular Correspondence, Peru. Vol. 260 (1870).

Archives des Affaires Etrangères de Paris (AAEP)

—Correspondance Commerciale des Consuls. Lima, Islay.

—Correspondance Commerciale et Consulaire. Vol. 1 (1828, 1830, 1846).

Periodicals

Anales de la Escuela de Construcciones Civiles y de Minas. Lima, 1880.

Anales de la Escuela de Ingenieros de Construcciones Civiles, de Minas e Industrias. Lima, 1901.

Boletín del Cuerpo de Ingenieros de Minas del Perú. Lima, 1901, 1902, 1903, 1904, 1905, 1916, 1917, 1918.

Boletín de minas, industria y construcciones. Lima, 1887, 1890, 1891, 1895, 1897.

El comercio. Lima, 1839, 1855, 1856, 1857, 1858, 1859, 1869.

El conciliador. Lima, 1830.

Memorial de ciencias naturales y de industria nacional y extranjera. Lima, 1828.

Mercurio peruano. Organo de la Sociedad de Amantes del País, vol. 1 (Lima), January 1791 (facsimile reproduction by the Biblioteca Nacional, Lima).

La prensa. Lima, 1910.

Articles, Pamphlets, and Books

Alayza, Oscar. *La industria minera en el Perú, 1936*. Lima: Ministerio de Fomento, 1937.

"Apuntes sobre las minas de carbón del Perú." *Boletín de minas, industria, y construcciones* 13, 13 (Lima, 30 March 1897).

Arana, Ricardo, ed. *Colección de leyes, decretos, y resoluciones que forman la legislación de minas del Perú.* Lima: n.p., n.d.

Arona, Juan de. *La inmigración en el Perú: Monografía histórico-crítica.* Lima: Imprenta del Universo, 1891.

Barba, Alvaro Alonso. *Arte de los metales.* Potosí: Colección de la Cultura Boliviana, 1968. Original edition, Madrid, 1640.

Bueno, Ismael C.. "Asiento del Cerro de Pasco." *Boletín de minas, industria, y construcciones* 3, 3 (Escuela Especial de Ingenieros de Lima, 1887).

———. "Asiento del Cerro de Pasco." *Boletín de minas, industria, y construcciones* 7, 7 (Escuela Especial de Ingenieros de Lima, 1891).

———. "Informe sobre el asiento mineral del Cerro de Pasco," in *Boletín de minas, industria y construcciones* (Lima: Escuela Especial de Ingenieros), year 8, vol. 8 (1892).

Castro Pozo, Hildebrando. *Nuestra comunidad indígena.* Lima: Perugraph Editores, 1979. Original edition, 1924.

Colección Documental de la Independencia del Perú. *Misiones peruanas, 1820–1826.* Vol. 11. Lima: CDIP, 1975.

———. *Misión García del Rio-Paroissien.* Tome 11, vol. 2. Lima: CDIP, 1973.

Concolorcorvo (Alonso Carrió de la Vandera). *Itinéraire de Buenos Aires à Lima.* Translated from the Spanish by Yvette Billod. Paris: Institute des Hautes Etudes de l'Amérique Latine, 1961. Original edition, Lima: n.p., 1776.

Costa y Laurent, Federico. *Reseña histórica de los ferrocarriles del Perú.* Lima: Litografía y Tipografía Carlos Fabri, 1908.

Dancuart, Emilio, and J. M. Rodríguez. *Anales de la hacienda pública del Perú (1821–1889).* 19 vols. Lima: Imprenta de La Revista, 1902–1926.

Dávalos y Lissón, Pedro. "La industria minera." In *El Perú.* Lima, n.p., 1900.

———. *La primera centuria: Causas geográficas, políticas, y económicas que han detenido el progreso moral y material del Perú en el primer siglo de su vida independiente.* 4 vols. Lima: Librería e Imprenta Gil, 1919–1926.

Davelouis, H. *Informe que el que suscribe eleva á la consideración de los poderes legislativo y ejecutivo sobre el estado actual de la minería en el Perú.* Lima: Imprenta de Huerta y Compañía, 1863.

Denegri, Marco Aurelio. *La crisis del enganche.* Lima: San Martín y Compañía, 1911.

Dirección de Estadística. *Censo general de la república del Perú formado en 1876.* 7 vols. Lima: Dirección de Estadística, 1878.

———. *Estadística de las minas de la república del Perú en 1878.* Lima: Imprenta del Estado, 1879.

Du Chatenet, Maurice. *Estado actual de la industria minera en el Cerro de Pasco.* Lima: Anales de la Escuela de Construcciones Civiles y de Minas, 1880.

Escuela Especial de Ingenieros de Lima. "Las minas de Bolivia." *Boletín de minas, industrias y construcciones* 6 (Lima, 1890).

Esteves, Luis. *Apuntes para la historia económica del Perú.* Lima: Centro de Estudios de Población y Desarrollo, 1971. Original edition, Lima: Imprenta Huallaga, 1882.

Fort, Michel. "Asiento mineral del Cerro de Pasco." *Anales de la escuela de ingenieros de construcciones civiles, de minas e industrias.* Lima, 1901, 8–16.

Fuchs, F. C. "Mineral de Vinchos y oficina de Humanrauca." *Boletín de minas, industria, y construcciones.* Escuela Especial de Ingenieros de Lima 11 (1895).

Fuentes Castro, Paulino, ed. *Nueva legislación peruana.* Lima: Editor de El Diario Judicial, 1903.

García Calderón, Francisco. *Diccionario de la legislación peruana.* Paris: Librería de Laroque, 1879.

Garcia Llanos. *Diccionario y maneras de hablar que se usan en las minas y sus labores en los ingenios y beneficios de los metales.* Ed. Thierry Saignes and Gunnar Mendoza. La Paz: Museo Nacional de Etnografía y Folklore, 1983. Original edition, 1609.

Garland, Alejandro. *La industria del petróleo en el Perú en 1901.* Boletín no. 2, Cuerpo de Ingenieros de Minas, Lima, 1902.

Habich, Eduardo A. V. de. "Código de minería." *Anales de la Escuela de Construcciones Civiles y de Minas del Perú*, vol. 3 (Lima, 1883).

———. *Yacimientos carboníferos del distrito de Checras*. Lima: Cuerpo de Ingenieros de Minas, 1904.

Herrmann, A. *La producción en Chile de los metales i minerales desde la conquista hasta fines del año 1902*. Santiago: n.p., 1903.

Humphreys, R. A., ed. *British Consular Reports on the Trade and Politics of Latin America, 1824–1826*. London: Royal Historical Society, 1940.

Jiménez, Carlos P. *Estadística minera en 1915*. Lima: Cuerpo de Ingenieros de Minas del Perú, 1916.

———. *Estadística minera en 1916*. Lima: Cuerpo de Ingenieros de Minas, 1917.

———. "Reseña histórica de la minería en el Perú." In *Síntesis de la minería peruana en el centenario de Ayacucho*, ed. Dirección de Minas y Petrole. Lima: Imprenta Torres Aguirre, 1924.

Lafond, Gabriel. "Impresiones de Lima." In *El Perú visto por viajeros*, ed. Estuardo Núñez, 1: 102–9. 2 vols. Lima: Ediciones Peisa, 1973. Original edition, n.p., 1822.

Málaga Santolalla, Fermín. *Recursos minerales de la provincia de Hualgayoc*. Lima: Cuerpo de Ingenieros de Minas, 1902.

———. *Los yacimientos carboníferos de la provincia de Celendín*. Lima: Cuerpo de Ingenieros de Minas, 1905.

Manning, William R., ed. *Diplomatic Correspondence of the United States: Inter-American Affairs, 1831–1860*. Washington, D.C.: Carnegie Endowment for International Peace, 1938.

———. *Diplomatic Correspondence of the United States Concerning the Independence of the Latin American Nations*. New York: Oxford University Press, 1925.

Martinet, J. B. H. *La agricultura en el Perú*. Lima: Centro Peruano de Historia Económica, Universidad Nacional Mayor de San Marcos, 1977. Original edition, 1877.

———. *Carestía de víveres en Lima*. Lima: Centro Peruano de Historia Económica, Universidad Nacional Mayor de San Marcos, 1977. Original edition, 1875.

McCulloch, J. M. *A Dictionary, Geographical, Statistical, and Historical of the Various Countries, Places, and Principal Natural Objects in the World*. 2 vols. London: Brown, Green and Longmans Publishers, 1846.

Ministro de Gobierno, policía, y obras públicas: Memoria que presenta al Congreso Ordinario de 1874. Lima: Imprenta de El Comercio, 1874.

Ministerio de Hacienda. *Padrón general de minas de 1887*. Lima: Imprenta del Estado, 1887.

Ministerio de Hacienda y Comercio. *Extracto estadístico del Perú*. Lima: Ministerio de Hacienda y Comercio, 1931–1933.

Morales y Ugalde, José. *Manifestación del estado de la hacienda de la república del Perú en fin de Abril de 1827: Presentada al soberano congreso constituyente por el ciudadano encargado de la dirección del ministerio*. Lima: Imprenta Rep. por J. M. Concha, 1827.

Nieto, Juan Crissóstomo, and Mariano Santos de Quirós. *Colección de leyes, decretos, y ordenes publicadas en el Perú desde su independencia*. 13 vols. Lima: Imprenta de la Colección, 1864.

Olaechea, Teodorico. *Apuntes sobre la minería en el Perú*. Lima: Imprenta de la Escuela de Ingenieros, 1898.

Ortiz de Zevallos, Carlos, ed. *Las primeras misiones diplomáticas en América*. Lima: Colección Documental de la Independencia del Perú, 1975.

Ministerio de Hacienda. *Padrón general de minas correspondiente al segundo semestre del año de 1878*. Lima: Imprenta del Estado, 1878.

Paz Soldán, Mariano Felipe. *Diccionario geográfico-estadístico del Perú*. Lima: Imprenta del Estado, 1877.

Paz Soldán, Mateo. *Geografía del Perú*. Paris: Librería de Fermin Didot, Hermanos, Hijos y Cia., 1862.

Pflucker, Carlos Renardo. *Exposición que presenta al supremo gobierno con motivo de las ultimas ocurrencias acaecidas en la hacienda mineral de Morococha*. Lima: Imprenta del Correo Peruano, 1846.

Raimondi, Antonio. "Memoria sobre el Cerro de Pasco y la montaña de Chanchamayo." In *El Perú*. Lima: Editores Técnicos Asociados, 1965, 4:444–88. Original edition, Lima: Imprenta del Estado, 1885.

———. *El Perú*. Vol. 4. 2d ed. Lima: Librería e Imprenta Gil, 1902. Original edition, Lima: Imprenta del Estado, 1885.

————. *El Perú.* Facsimile edition, Lima: Editores Técnicos Asociados, 1965–1966. Original edition, 6 vols., Lima: Sociedad Geográfica de Lima and Imprenta del Estado, 1874–1913.

Rivero y Ustáriz, Mariano Eduardo de. *Colección de memorias científicas, agrícolas, e industriales publicadas en distintas épocas.* 2 vols. Brussels: Imprenta de H. Goemaere, 1857.

Rodríguez, José Manuel. *Estudios económicos y financieros y ojeada sobre la hacienda pública del Perú y la necesidad de su reforma.* Lima: Librería Gil, 1895.

Soetbeer, Adolph. *Edelmetall-Produktion.* Gotha: Justus Perthes, 1879.

Stiglich, Germán. *Diccionario geográfico del Perú.* 2 vols. Lima: Imprenta Torres Aguirre, 1922–1923 (second edition).

Superintendencia de Aduanas. *Estadística general del comercio exterior del Perú.* Lima, n.p. 1897–1900.

Tschudi, Johann Jakob von. *Testimonio del Perú, 1838–1842.* Lima: Consejo Consultivo Suiza-Perú, 1966.

Ulloa, Antonio de, and Jorge Juan y Santacilia de Ulloa. *Relación histórica del viaje a la América meridional.* 2 vols. Madrid: Fundación Universitaria Española, 1989. Original edition, Madrid: n.p., 1748.

Venturo, Pedro C. "Excursiones científicas: Viaje al asiento mineral del Cerro de Pasco." *Boletín de minas, industria, y construcciones* 13, 7 (Lima, 10 August 1897): 38–59.

Yáñez León, Juan M. *Yacimientos carboníferos de las provincias de Pallasca, Huaylas y Yungay.* Lima: Cuerpo de Ingenieros de Minas del Perú, 1918.

Zulen, Pedro. "El enganche de indios." *La prensa* (Lima), 7 October 1910.

Secondary Sources

Aguirre, Carlos. *Agentes de su propia libertad: Los esclavos y la desintegración de la esclavitud, 1821–1854.* Lima: Pontificia Universidad Católica del Perú, 1993.

Aguirre, Carlos, and Charles Walker, eds. *Bandoleros, abigeos, y montoneros: Criminalidad y violencia en el Perú, siglos dieciocho-veinte.* Lima: Instituto de Apoyo Agrario, 1990.

Alberti, Giorgio, and Enrique Mayer, eds. *Reciprocidad e intercambio en los Andes peruanos*. Lima: Instituto de Estudios Peruanos, 1974.

Alcalde Mongrut, Arturo. *Mariano de Rivero-Federico Villareal*. Lima: Editorial Universitaria, 1966.

Allpanchis. *Arrieros y circuitos mercantiles andinos*. Vol. 18, no. 21. Cusco: Instituto de Pastoral Andina, 1983.

———. *Conflicto y campesinado en la minería andina*. Vol. 22. Cusco: Instituto de Pastoral Andina, 1985.

Anderson, Anthony B., Peter H. May, and Michael J. Balick. *The Subsidy from Nature: Palm Forests, Peasantry, and Development on an Amazon Frontier*. New York: Columbia University Press, 1991.

Appleby, Gordon. "Exportation and Its Aftermath: The Spatial Economic Evolution of the Regional Marketing System in Highland Puno, Peru." Ph.D. dissertation, Stanford University, 1978.

———. "Las transformaciones del sistema de mercados en Puno, 1890–1960." *Análisis* 8–9 (Cuadernos de Investigación, Lima, May–December 1979): 55–71.

Arduz Eguía, Gastón. *Ensayos sobre la historia de la minería alto-peruana*. Madrid: Editorial Paraninfo, 1985.

Arguedas, José María. "Evolución de las comunidades indígenas: El valle del Mantaro y la ciudad de Huancayo: un caso de fusión de culturas no comprometidas por la acción de las instituciones de origen colonial." *Revista del museo nacional* 26 (Lima, 1957).

———. *Indios, mestizos, y señores*. Lima: Editorial Horizonte, 1985.

Aristotle. *Metaphysics*. Trans. Richard Hope. Ann Arbor: University of Michigan Press, 1960.

Ashton, T. S. *The Industrial Revolution, 1760–1830*. Oxford: Oxford University Press, 1948.

Assadourian, Carlos Sempat. "Modos de producción, capitalismo, y subdesarrollo en América Latina." In *Modos de Producción en América Latina*, ed. Carlos Sempat Assadourian et al., 47–81. Cuadernos de Pasado y Presente no. 40, Cordoba, 1973.

———. "La producción de la mercancía dinero en la formación del mercado interno colonial: El caso del espacio peruano, siglo dieciseis." In

Ensayos sobre el Desarrollo Económico de México y América Latina (1500–1975), ed. Enrique Florescano, 223–92. Mexico City: Fondo de Cultura Económica, 1979.

———. *El sistema de la economía colonial: Mercado interno, regiones, y espacio económico.* Lima: Instituto de Estudios Peruanos, 1982.

Assadourian, Carlos Sempat, Heraclio Bonilla, Antonio Mitre, and Tristan Platt. *Minería y espacio económico en los Andes, siglos deiciséis-veinte.* Lima: Instituto de Estudios Peruanos, 1980.

Assadourian, Carlos Sempat, C.F.S. Cardaso et al. *Modos de producción en América Latina.* Cuadernos de Pasado y Presente no. 40, Cordoba, 1973.

Avendaño Hübner, Jorge. *Miraflores de antaño.* Lima: Universidad Peruana Cayetano Heredia, 1989.

Avila, Dolores, Inés Herrera, and Rina Ortiz, eds. *Empresarios y política minera.* Mexico City: Instituto Nacional de Antropología e Historia, 1992.

Bairoch, Paul. *Commerce extérieur et développement économique de l'Europe au dix-neuvième siècle.* Paris: Mouton and Ecole des Hautes Etudes en Sciences Sociales, 1976.

Bakewell, Peter. *Miners of the Red Mountain: Indian Labor in Potosí, 1545–1650.* Albuquerque: University of New Mexico Press, 1984.

———. "Mining." In *Colonial Spanish America,* ed. Leslie Bethell, 203–49. Cambridge: Cambridge University Press, 1988.

Ballantyne, Janet Campbell. "The Political Economy of the Peruvian 'Gran Minería.'" Ph.D. dissertation, Cornell University, 1976.

Bargalló, Modesto. *La minería y metalurgia en la América española durante la época colonial.* Mexico City: Fondo de Cultura Económica, 1955.

Barrantes, Salvador, and Nora Velarde. *El capital internacional en la sierra central.* Lima: Universidad Nacional Federico Villareal, 1983.

Basadre, Jorge. *Chile, Perú, y Bolivia independientes.* Barcelona: Salvat Editores, 1948.

———. *Historia de la república del Perú, 1822–1933.* 6th ed. 17 vols. Lima: Editorial Universitaria, 1968–70.

———. *Introducción a las bases documentales para la historia de la república del Perú con algunas reflexiones.* 2 vols. Lima: Ediciones P. L. Villanueva, 1971.

———. *La multitud, la ciudad, y el campo en la historia del Perú.* Lima: Ediciones Treintaitrés and Mosca Azul Editores, 1980. Original edition, n.p., 1929.

———. *Perú: Problema y posibilidad.* Lima: Banco Internacional del Perú, 1978. Original edition, n.p., 1931.

Becker, Marjorie. *Setting the Virgin on Fire: Lázaro Cárdenas, Michoacán Peasants, and the Redemption of the Mexican Revolution.* Berkeley: University of California Press, 1995.

Bethell, Leslie, ed. *Colonial Spanish America.* Cambridge: Cambridge University Press, 1988.

Bieber, Leon E. "Empresarios mineros en el siglo diez y nueve: Bases para su caracterización social." Cochabamba, Bolivia: Instituto de Estudios Sociales y Económicos, 1980 (mimeo).

Blanchard, Peter. *Slavery and Abolition in Early Republican Peru.* Wilmington, Del.: Scholarly Resources, 1992.

Bloch, Marc. *The Historian's Craft.* Trans. Peter Putnam. New York: Vintage Books, 1953.

Boloña, Carlos. "Perú: Estimaciones preliminares del producto nacional, 1900–1942." *Apuntes: Revista de ciencias sociales* 13 (Lima, Universidad del Pacífico, 1983): 3–13.

———. "Tariff Policies in Peru, 1880–1980." Ph.D. dissertation, Oxford University, 1981.

Bonilla, Heraclio. "Aspects de l'histoire économique et sociale du Pérou au dix-neuvième siècle." 2 vols. Thèse de doctorat du troisième cycle, Université de Paris, 1970.

———. "La coyuntura comercial del siglo deicinueve en el Perú." *Revista del museo nacional* 35 (Lima, 1967–68): 159–87.

———. *Guano y burguesía en el Perú.* Lima: Instituto de Estudios Peruanos, 1974. 2d ed., 1984.

———. "El impacto de los ferrocarriles: Algunas proposiciones." *Historia y cultura* (Revista del Museo Nacional de Historia, Lima, 1972): 93–120.

————. *Los mecanismos de un control económico.* Vol. 5 of Gran Bretaña y el Perú. Lima: Instituto de Estudios Peruanos and Fondo del Libro del Banco Industrial, 1977.

————. *El minero de los Andes: Una aproximación a su estudio.* Lima: Instituto de Estudios Peruanos, 1974.

————. "The New Profile of Peruvian History." *Latin American Research Review* 16, 3 (1981): 210–24.

————. *Un siglo a la deriva: Ensayos sobre el Perú, Bolivia, y la guerra.* Lima: Instituto de Estudios Peruanos, 1980.

————. "The War of the Pacific and the National and Colonial Problem in Peru." *Past and Present,* no. 81 (November 1978): 92–118.

————, ed. *Las crisis económicas en la historia del Perú.* Lima: Centro Latinoamericano de Historia Económica y Social and Fundación Friedrich Ebert, 1986.

————, ed. *Gran Bretaña y el Perú: Informes de los cónsules británicos, 1828–1919.* 4 vols. Lima: Instituto de Estudios Peruanos and Fondo del Libro del Banco Industrial, 1975.

————, ed. *El sistema colonial en la América española.* Barcelona: Editorial Crítica, 1991.

Bonilla, Heraclio, Lía del Río, and Pilar Ortíz de Zevallos. "Comercio libre y crisis de la economía andina: El caso del Cuzco." *Histórica* 2, 1 (Lima, Pontificia Universidad Católica del Perú, July 1978): 1–25.

Bowser, Frederick P. *The African Slave in Colonial Peru, 1524–1650.* Stanford: Stanford University Press, 1974.

Brading, David A. "Bourbon Spain and its American Empire." In *Colonial Spanish America,* ed. Leslie Bethell, 112–62. Cambridge: Cambridge University Press, 1988.

————. *Miners and Merchants in Bourbon Mexico.* Cambridge: Cambridge University Press, 1970.

Brading, David A., and Harry Cross. "Colonial Silver Mining: Mexico and Peru." *Hispanic American Historical Review* 52, 4 (November 1972): 545–79.

Bratter, Herbert. *The Silver Market.* Washington, D.C.: Government Printing Office, 1932.

Braudel, Fernand. *La dynamique du capitalisme*. Paris: Flammarion, 1985.

———. *Ecrits sur l'histoire*. Paris: Flammarion, 1969.

———. *The Mediterranean and the Mediterranean World in the Age of Philip II*. Trans. from the French by Siân Reynolds. New York: Harper and Row, 1976. Original edition, Paris: Colin, 1949.

———. *On History*. Trans. Sarah Matthews. Chicago: University of Chicago Press, 1980.

Braudel, Fernand, and Ernest Labrousse, eds. *Histoire économique et sociale de la France*. Paris: Presses Universitaires de France, 1976.

Bravo Bresani, Jorge, ed. *La oligarquía en el Perú*. Lima: Instituto de Estudios Peruanos, 1971.

Bronner, Fred. "Peruvian Historians Today: Historical Setting." *Americas* 43, 3 (1986): 245–77.

Burga, Manuel. *De la encomienda a la hacienda capitalista: El valle del Jequetepeque del siglo deiciséis al veinte*. Lima: Instituto de Estudios Peruanos, 1976.

Burga, Manuel, and Wilson Reátegui. *Lanas y capital mercantil en el sur: La Casa Ricketts, 1895–1935*. Lima: Instituto de Estudios Peruanos, 1981.

Caballero Martín, Víctor. *Imperialismo y campesinado en la sierra central*. Huancayo, Peru: Instituto de Estudios Andinos, 1981.

Cajías, Fernando. *La provincia de Atacama, 1825–1842*. La Paz, 1975.

Camprubí, Carlos. *El banco de la emancipación*. Lima: Imprenta P.L. Villanueva, 1960.

———. *Bancos de rescate, 1821–1832*. Lima, n.p., 1963.

———. *Historia de los bancos en el Perú (1860–1879)*. Lima: Talleres Gráficos de la Editorial Lumen, 1957.

Cardoso, Fernando Henrique. "The Consumption of Dependency Theory in the United States." *Latin American Research Review* 12, 3 (1977): 7–24.

Carmagnani, Marcello. *Les mécanismes de la vie économique dans une société coloniale: Le Chili, 1680–1830*. Paris: SEVPEN, 1973.

Carrión Ordóñez, Enrique. "Fuentes bibliográficas sobre los idiomas del Perú." *Humanidades* 5 (Pontificia Universidad Católica del Perú, Lima, 1972–1973): 113–29.

Casaverde, Juvenal, ed. *Campesinado y capitalismo*. Huancayo, Peru: Instituto de Estudios Andinos, 1979.

Centro de Estudios de Población y Desarrollo. *Informe demográfico del Perú*. Lima: CEPD, 1972.

Chaca, Pablo. *Capitalismo minero*. Lima: Universidad Nacional Mayor de San Marcos, 1980.

Chaianov, Alexander V. *The Theory of Peasant Economy*. Madison: University of Wisconsin Press, 1986. Original Russian edition, 1923.

Chaunu, Pierre. *Histoire quantitative, histoire sérielle*. Paris: Cahiers des Annales, 1978.

———. *Séville et l'Atlantique (1504–1650): Structures et conjoncture de l'Atlantique espagnol et hispano-américain*. 12 vols. Paris: SEVPEN, 1955–1960.

Chocano, Magdalena. "Circuitos comerciales y auge minero en la sierra central a fines de la época colonial." *Allpanchis* 18, 21 (Instituto de Pastoral Andina, Cusco, 1983): 3–26.

———. "Comercio en Cerro de Pasco a fines de la época colonial." Tesis de historia, Pontificia Universidad Católica del Perú, Lima, 1982.

Cipolletti, María Susana. "Llamas y mulas, trueque y venta: El testimonio de un arriero puneño." *Revista Andina* 4 (Centro Bartolomé de las Casas, Cusco, 1984): 513–38.

Clayton, Lawrence A. *Grace: W. R. Grace and Co., The Formative Years, 1850–1930*. Ottawa, Ill.: Jameson Books, 1985.

Coatsworth, John H. *Growth against Development: The Economic Impact of Railroads in Porfirian Mexico*. De Kalb: Northern Illinois University Press, 1981.

Cole, Jeffrey A. *The Potosí Mita, 1573–1700: Compulsory Indian Labor in the Andes*. Stanford: Stanford University Press, 1985.

Contreras, Carlos. "Minería y población en los Andes: Cerro de Pasco en el siglo diecinueve." Research report, Instituto de Estudios Peruanos, Lima, September 1984 (manuscript).

———. "Mineros, arrieros, y ferrocarril en Cerro de Pasco, 1870–1904." *Hisla*, no. 4 (Revista Latinoamericana de Historia Económica y Social, Lima, 1984): 3–20.

――――. *Mineros y campesinos en los Andes: Mercado laboral y economía campesina en la sierra central, siglo diecinueve.* Lima: Instituto de Estudios Peruanos, 1988.

Corbin, Alain. *The Lure of the Sea: The Discovery of the Seaside in the Western World, 1750–1840.* Translated from the French by Jocelyn Phelps. Berkeley: University of California Press, 1994.

Cotlear, Daniel. "Enganche, salarios, y mercado de trabajo en la ceja de selva peruana." *Análisis* 7 (Cuadernos de Investigación, Lima, January-April 1979): 67–85.

――――. "El sistema de enganche a principios del siglo veinte: Una versión diferente." Tesis de economía, Lima: Pontificia Universidad Católica del Perú, 1979.

Cotler, Julio. *Clases, estado, y nación en el Perú.* Lima: Instituto de Estudios Peruanos, 1978.

Dargent C., Eduardo. *El billete en el Perú.* Lima: Banco Central de Reserva del Perú, 1979.

Dávila, Dilma. "Talara, los petroleros, y la huelga de 1931." Tesis de sociología, Pontificia Universidad Católica del Perú, Lima, 1976.

Degregori, Carlos Iván. *Ayacucho 1969–1979: El surgimiento de Sendero Luminoso.* Lima: Instituto de Estudios Peruanos, 1990.

――――. *Ayacucho, raíces de una crisis.* Ayacucho: Instituto de Estudios Regionales José María Arguedas, 1986.

Del Busto Duthurburu, José Antonio. *Historia y leyenda del viejo Barranco.* Lima: Editorial Lumen, 1985.

Derpich, Vilma. "Introducción al estudio del trabajador coolie en el Perú del siglo diecinueve." Tesis de historia, Universidad Nacional Mayor de San Marcos, Lima, 1976.

De Secada, C. Alexander G. "Armas, guano, y comercio marítimo: Los intereses de W. R. Grace en el Perú, 1865–1885." *Hisla,* no. 7 (Revista Latinoamericana de Historia Económica y Social, Lima, 1986): 105–29.

Deustua, José. "El ciclo interno de la producción del oro en el tránsito de la economía colonial a la republicana: Perú, 1800–1840." *Hisla,* no. 3 (Revista Latinoamericana de Historia Económica y Social, Lima, 1984): 23–49.

———. "De la minería a la acuñación de moneda y el sistema monetario en el Perú del siglo diecinueve." In *Apuntes sobre el Proceso Histórico de la Moneda. Perú, 1820–1920,* ed. Javier Ramírez Gastón and Soledad Arispe. Lima: Banco Central de Reserva del Perú, 1993.

———. "Derroteros de la etnohistoria en el Perú." *Allpanchis* 14–15 (Instituto de Pastoral Andina, Cusco, 1980): 173–78.

———. "La minería, las clases sociales, y la independencia del Perú." *Análisis* 12 (Cuadernos de Investigación, Lima, 1983): 50–62.

———. *La minería peruana y la iniciación de la república, 1820–1840.* Lima: Instituto de Estudios Peruanos, 1986.

———. "Mines, monnaie, et hommes dans les Andes: Une histoire économique et sociales de l'activité minière dans le Pérou du dix-neuvième siècle." 2 vols. Thèse de doctorat (Paris: Ecole des Hautes Etudes en Sciences Sociales, 1989).

———. "Mining Markets, Peasants, and Power in Nineteenth-Century Peru." *Latin American Research Review* 29, 1 (1994): 29–54.

———. "Producción minera y circulación monetaria en una economía andina: El Perú del siglo diecinueve." *Revista Andina* 4, 2 (Cusco, 1986): 319–54.

———. "Routes, Roads, and Silver Trade in Cerro de Pasco, 1820–1860: The Internal Market in Nineteenth-Century Peru." *Hispanic American Historical Review* 74, 1 (1994): 1–31.

———. "Sobre movimientos campesinos e historia regional en el Perú moderno: Un comentario bibliográfico." *Revista Andina* 1, 1 (Cusco, 1983): 219–40.

———. "The Socavon of Quiulacocha and the Steam Engine Company: Technology and Capital Investment in Cerro de Pasco, 1820–1840." In *Region and Class in Modern Peruvian History,* ed. Rory Miller, 35–75. Liverpool: University of Liverpool, 1987.

Deustua, José, and José Luis Rénique. *Intelectuales, indigenismo, y descentralismo en el Perú, 1897–1931.* Cusco: Centro de Estudios Rurales Andinos Bartolomé de las Casas, 1984.

Deustua Pimentel, Carlos. "La minería peruana en el siglo dieciocho (aspectos de un estudio entre 1790 y 1796)." *Humanidades,* no. 3 (Pontificia Universidad Católica del Perú, Lima, 1969): 29–47.

DeWind, André. "Peasants Become Miners: The Evolution of Industrial Mining Systems in Peru." Ph.D. dissertation, Columbia University, 1977.

Deyon, Pierre. "L'enjeu des discussions autour du concept de 'protoindustrialisation.'" *Revue du Nord* 61, 240 (1979): 9–18.

Díaz, Alida. "El censo general de 1876 en el Perú." Lima: Seminario de Historia Rural Andina, Universidad Nacional Mayor de San Marcos, 1974 (mimeo).

Dirección General de Minería. *Anuario de la minería del Perú.* Lima: Ministerio de Energía y Minas, 1969.

Dobyns, Henry, and Paul L. Doughty. *Peru, a Cultural History.* New York: Oxford University Press, 1976.

Dollfus, Olivier. "Les Andes intertropicales: Une mosaique changeante." *Annales: Economies, sociétés, civilizations* 33, 5–6 (Paris, September–December 1978): 895–905.

———. *Le Pérou: Introduction géographique à l'étude du développement.* Paris: Institut des Hautes Etudes de l'Amérique Latine, 1968.

———. *El reto del espacio andino.* Lima: Instituto de Estudios Peruanos, 1981.

Dopsch, Alphons. *Economía natural y economía monetaria.* Mexico City: Fondo de Cultura Económica, 1943. Original German edition, 1930.

Dore, Elizabeth. "Accumulation and Crisis in Peruvian Mining: 1900–1977." Ph.D. dissertation, Columbia University, 1980.

ECO. *Crisis minera y sobre-explotación de la fuerza de trabajo.* Lima: ECO, Grupo de Investigaciones Económicas, 1980.

Eco, Umberto. *The Open Work.* Trans. Anna Cancogni. Cambridge, Mass.: Harvard University Press, 1989.

Eco, Umberto, Richard Rorty, Jonathan Culler, and Christine Brooke-Rose. *Interpretation and Overinterpretation,* ed. Stefan Collini. Cambridge: Cambridge University Press, 1992.

Espinosa Bravo, Clodoaldo Alberto. *El hombre de Junín frente a su paisaje i a su folklore.* 2 vols. Lima: Talleres Gráficos P. L. Villanueva, 1967.

Espinoza Claudio, César, and José Boza Monteverde. "Alcabalas y protesta popular: Cerro de Pasco 1780." Lima: Universidad Nacional Mayor de San Marcos, 1981 (mimeo).

Espinoza Soriano, Waldemar. *Enciclopedia departamental de Junín.* Huancayo: Editor Enrique Chipoco Tovar, 1973.

Fioravanti, Eduardo. *Latifundismo y sindicalismo agrario en el Perú.* Lima: Instituto de Estudios Peruanos, 1974.

Fisher, John R. *Government and Society in Colonial Peru: The Intendant System, 1784–1814.* London: Athlone Press, 1970.

———, ed. "Matrícula de los mineros del Perú, 1790." Lima: Seminario de Historia Rural Andina, Universidad Nacional Mayor de San Marcos, 1975 (mimeo).

———. *Minas y mineros en el Perú Colonial, 1776–1824.* Lima: Instituto de Estudios Peruanos, 1977.

———. "Mineros y minería de plata en el virreinato del Perú, 1776–1824." *Histórica* 3, 2 (Pontificia Universidad Católica del Perú, Lima, 1979): 57–70.

———. *Silver Mines and Silver Miners in Colonial Peru.* Liverpool: University of Liverpool, 1977.

Florescano, Enrique, ed. *Ensayos sobre el desarrollo económico de México y América Latina (1500–1975).* Mexico City: Fondo de Cultura Económica, 1979.

Flores Galindo, Alberto. *La agonía de Mariátegui.* Lima: Centro de Estudias y Promoción del Desarrollo, 1980.

———. *Arequipa y el sur andino, siglos dieciocho-veinte.* Lima: Editorial Horizonte, 1977.

———. *Aristocracia y plebe: Lima, 1760–1830.* Lima: Mosca Azul Editores, 1984.

———. *Buscando un Inca: Identidad y utopía en los Andes.* Havana: Casa de las Américas, 1986.

———. *Los mineros de la Cerro de Pasco, 1900–1930.* Lima: Pontificia Universidad Católica del Perú, 1974.

———. "La pesca y los pescadores en la costa central (siglo dieciocho)." *Histórica* 5, 2 (Pontificia Universidad Católica del Perú, Lima, December 1981): 159–65.

Flores Ochoa, Jorge. *Los pastores de Paratía.* Mexico City: Instituto Indigenista Interamericano, 1968.

————, ed. *Pastores de puna. [Uywamichiq punarunakuna.]* Lima: Instituto de Estudios Peruanos, 1977.

Fundación Rio Tinto. *La comarca de Rio Tinto: Un territorio de mina.* Huelva, Spain: Fundación Rio Tinto, 1994.

Gallagher, J., and R. Robinson. "The Imperialism of Free Trade." *Economic History Review*, 2d series 6 (1953): 1–15.

Garcilaso de la Vega. *Comentarios reales de los Incas.* 3 vols. Lima: Ediciones Peisa, 1973.

Gardiner, C. Harvey. *The Japanese and Peru, 1873–1973.* Albuquerque: University of New Mexico Press, 1975.

Garraty, John. *A Short History of the American Nation.* New York: Harper and Row, 1981.

Giesecke, Margarita. *Masas urbanas y rebelión en la Historia: Golpe de Estado, Lima, 1872.* Lima: Centro de Estudios y Divulgación de Historia Popular, 1978.

Glave, Luis Miguel. *Trajinantes: Caminos indígenas en la sociedad colonial, siglos dieciséis– diecisiete.* Lima: Instituto de Apoyo Agrario, 1989.

————. *Vida, símbolos, y batallas: Creación y recreación de la comunidad indígena: Cusco, siglos dieciséis-veinte.* Mexico City: Fondo de Cultura Económica, 1992.

Golte, Jürgen. *Repartos y rebeliones: Tupac Amaru y las contradicciones de la economía colonial.* Lima: Instituto de Estudios Peruanos, 1980.

Gonzales, Michael J. *Plantation Agriculture and Social Control in Northern Peru, 1875–1933.* Austin: University of Texas Press, 1985.

Gootenberg, Paul. *Between Silver and Guano: Commercial Policy and the State in Postindependence Peru.* Princeton: Princeton University Press, 1989.

————. "*Carneros y chuño*: Price Levels in Nineteenth-Century Peru." *Hispanic American Historical Review* 70, 1 (1990): 1–56.

————. *Imagining Development. Economic Ideas in Peru's "Fictitious Prosperity" of Guano, 1840–1880.* Berkeley: University of California Press, 1993.

————. "Merchants, Foreigners, and the State: The Origins of Trade Policies in Post-Independence Peru." Ph.D. dissertation, University of Chicago, 1985.

————. "Population and Ethnicity in Early Republican Peru: Some Revisions." *Latin American Research Review* 26, 3 (1991): 109–57.

————. "The Social Origins of Protectionism and Free Trade in Nineteenth-Century Lima." *Journal of Latin American Studies* 14, 2 (1982): 329–58.

————. *Tejidos y harinas, corazones y mentes: El imperialismo norteamericano del libre comercio en el Perú, 1825–1840*. Lima: Instituto de Estudios Peruanos, 1989.

Goubert, Pierre. *Initiation à l'histoire de la France*. Paris: Fayard and Tallandier, 1984.

Gramsci, Antonio. *El materialismo histórico y la filosofía de Benedetto Croce*. Buenos Aires: Nueva Visión, 1972.

————. *Selections from the Prison Notebooks*. Ed. and trans. Q. Hoare and G. Nowell Smith. New York: International Publishers, 1971.

Guzmán, Augusto. *Historia de Bolivia*. Cochabamba: Editorial Los Amigos del Libro, 1990.

Hamilton, Earl J. *The American Treasure and the Price Revolution in Spain, 1501–1660*. Cambridge, Mass.: Harvard University Press, 1934.

Harris, Olivia, Brooke Larson, and Enrique Tandeter, eds. *La participación indígena en los mercados surandinos: Estrategias y reproducción social, siglos dieciséis al veinte*. La Paz: Centro de Estudios de la Realidad Social y Económica, 1988.

Hartz, Louis, ed. *The Founding of New Societies*. New York: Mentor Books, 1964.

Herrera Canales, Inés, Rina Ortiz Peralta, María Eugenia Romero Sotelo, and José Alfredo Uribe Salas. *Ensayos sobre minería mexicana, siglos dieciocho al veinte*. Mexico City: Instituto Nacional de Antropología e Historia, 1996.

Herrera Canales, Inés, and Rina Ortiz Peralta, eds. *Minería americana colonial y del siglo diecinueve*. Mexico City: Instituto Nacional de Antropología e Historia, 1994.

Hirschman, Albert O. *The Strategy of Economic Development*. New Haven: Yale University Press, 1958.

Hobsbawm, Eric J. *The Age of Capital, 1848–1875*. London: Abacus, 1977.

Hünefeldt, Christine. *Los Manuelos: Vida cotidiana de una familia negra en la Lima del siglo diecinueve: Una reflexión histórica sobre la esclavitud urbana.* Lima: Instituto de Estudios Peruanos, 1992.

———. *Paying the Price of Freedom: Family and Labor among Lima's Slaves, 1800-1854.* Berkeley: University of California Press, 1994.

———. "Viejos y nuevos temas de la historia económica del siglo diecinueve." In *Las Crisis Económicas en la Historia del Perú,* ed. Heraclio Bonilla, 33-60. Lima: Centro Latinoamericano de Historia Económica y Social and Fundación Friedrich Ebert, 1986.

Hunt, Shane. "Growth and Guano in Nineteenth-Century Peru." Discussion paper no. 34, Woodrow Wilson School, Princeton, 1973.

———. "Guano y crecimiento en el Perú del siglo diecinueve." *Hisla,* no. 4 (Revista Latinoamericana de Historia Económica y Social, Lima, 1984): 35-92.

———. *Price and Quantum Estimates of Peruvian Exports, 1830-1962.* Discussion paper no. 34, Woodrow Wilson School, Princeton, 1973.

Jacobsen, Nils. "Landtenure and Society in the Peruvian Altiplano: Azángaro Province, 1770-1920." Ph.D. dissertation, University of California, Berkeley, 1982.

———. *Mirages of Transition: The Peruvian Altiplano, 1780-1930.* Berkeley: University of California Press, 1993.

Jave, Noé, ed. *Jorge Basadre: La política y la historia.* Lima: Lluvia Editores, 1981.

Joseph, Gilbert M., and Daniel Nugent, eds. *Everyday Forms of State Formation: Revolution and the Negotiation of Rule in Modern Mexico.* Durham, N.C.: Duke University Press, 1994.

Kapsoli, Wilfredo. *Los movimientos campesinos en Cerro de Pasco, 1880-1963.* Huancayo, Peru: Instituto de Estudios Andinos, 1975.

———, ed. *Los movimientos campesinos en el Perú, 1879-1965.* Lima: Delva Editores, 1977.

Keith, Robert G. *Conquest and Agrarian Change: The Emergence of the Hacienda System on the Peruvian Coast.* Cambridge, Mass.: Harvard University Press, 1976.

Klarén, Peter F. *Modernization, Dislocation, and Aprismo: Origins of the*

Peruvian Aprista Party, 1870–1932. Austin: University of Texas Press, 1973.

Klein, Herbert S. *Bolivia: The Evolution of a Multi-Ethnic Society.* New York: Oxford University Press, 1982.

Klein, Philip A. "Economics: Allocation or Valuation?" In *The Economy as a System of Power,* ed. Warren J. Samuels, 7–33. New Brunswick, N.J.: Transaction Books, 1979.

Knight, Alan. "The Peculiarities of Mexican History: Mexico Compared to Latin America, 1821–1992." *Journal of Latin American Studies* 24 (quincentenary supplement, 1992): 99–144.

Kruijt, Dirk, and Menno Velinga. *Estado, clase obrera, y empresa transnacional: El caso de la minería peruana, 1900–1980.* Mexico City: Siglo Veintiuno Editores, 1983.

Labrousse, Ernest. *Esquisse du mouvement des prix et des revenus en France au dix-huitième siècle.* Paris: Dalloz, 1933.

LaCapra, Dominick. "History, Language, and Reading: Waiting for Crillon." *American Historical Review* 100, 3 (June 1995): 799–828.

Landes, David S. *The Unbound Prometheus: Technological Change and Industrial Development in Western Europe from 1750 to the Present.* Cambridge: Cambridge University Press, 1969.

Langer, Erick D. *Economic Change and Rural Resistance in Southern Bolivia, 1880–1930.* Stanford: Stanford University Press, 1989.

Langer, Erick D., and Gina L. Hames. "Commerce and Credit on the Periphery: Tarija Merchants, 1830–1914." *Hispanic American Historical Review* 74, 2 (May 1994): 285–316.

Langue, Frédérique, and Carmen Salazar-Soler. *Dictionnaire des termes miniers en usage en Amérique espagnole (seizième–dix-neuvième siècle).* Paris: Editions Recherche sur les Civilisations, 1993.

Lefevre, André. *Sous le Second Empire: Chemins de fer et politique.* Paris: Société d'Edition d'Enseignement Supérieur, 1951.

Lenin, Vladimir Ilich. *The Development of Capitalism in Russia.* Moscow: Progress Publishers, 1974. Original Russian edition, 1899.

Leong, Y. S. *Silver: An Analysis of Factors Affecting Its Price.* Washington, D.C.: Brookings Institution, 1933.

Levin, Jonathan. *The Export Economies: Their Pattern of Development in Historical Perspective.* Cambridge, Mass.: Harvard University Press, 1960.

Levine, Daniel H., ed. *Constructing Culture and Power in Latin America.* Ann Arbor: University of Michigan Press, 1993.

Llosa, Jorge Guillermo, ed. *Juan de Arona y la inmigración en el Perú.* Lima: Academia Diplomática del Perú, 1971.

Lomnitz, Larissa Adler, Claudio Lomnitz Adler, and Ilya Adler. "The Function of the Form: Power Play and Ritual in the 1988 Mexican Presidential Campaign." In *Constructing Culture and Power in Latin America,* ed. Daniel H. Levine, 357–401. Ann Arbor: University of Michigan Press, 1993.

López, Jacinto. *Manuel Pardo.* Lima: Editorial Gil, 1947.

López Soria, José Ignacio. "La escuela de ingenieros y la minería." In *Historia, problema, y promesa: Homenaje a Jorge Basadre,* ed. Francisco Miró Quesada, Franklin Pease, and David Sobrevilla, 2:149–69. Lima: Pontificia Universidad Católica del Perú, 1978.

———. *Historia de la Universidad Nacional de Ingeniería: Los años fundacionales, 1876–1909.* Lima: Universidad Nacional de Ingeniería, 1981.

Love, Joseph L., and Nils Jacobsen, eds. *Guiding the Invisible Hand: Economic Liberalism and the State in Latin American History.* New York: Praeger, 1988.

Lynch, John. *Bourbon Spain, 1700–1808.* Oxford: Basil Blackwell, 1989.

Macera, Pablo. "El arte mural cuzqueño, siglos dieciséis-veinte." *Apuntes 2,* 4 (Revista Semestral de Ciencias Sociales, Universidad del Pacífico, Lima, 1975).

———. "Arte y lucha social: Los murales de Ambaná (Bolivia)." *Allpanchis* 15, 17–18 (Instituto de Pastoral Andina, Cusco, 1981): 23–40.

———. "Estadísticas históricas del Perú: Sector minero (Precios)." Lima: Centro Peruano de Historia Económica, 1972 (mimeo).

———. "La historia en el Perú: Ciencia e ideología." *Amaru* 6 (Revista de Artes y Ciencias, Universidad Nacional de Ingeniería, Lima, April–June 1968).

————. "Mapas coloniales de haciendas cuzqueñas." Lima: Seminario de Historia Rural Andina de la Universidad Nacional Mayor de San Marcos, 1968 (mimeo).

————. *Pintores populares andinos.* Lima: Fondo del Libro del Banco de los Andes, 1979.

————. "Población rural en haciendas." Lima: Seminario de Historia Rural Andina, Universidad Nacional Mayor de San Marcos, 1976 (mimeo).

————. *Retablos andinos.* Lima: Universidad Nacional Mayor de San Marcos, 1981.

————. *Trabajos de historia.* 4 vols. Lima: Instituto Nacional de Cultura, 1977.

Macera, Pablo, Rosaura Andazabal, and Walter Carnero. *Los precios del Perú, siglos diecieéis-diecinueve: Fuentes.* 3 vols. Lima: Banco Central de Reserva del Perú, 1992.

Macera, Pablo, and Onorio Pinto. "Estadísticas históricas del Perú: Sector minero 2 (Volumen y valor)." Lima: Centro Peruano de Historia Económica, 1972 (mimeo).

Mallon, Florencia E. *The Defense of Community in Peru's Central Highlands: Peasant Struggle and Capitalist Transition, 1860–1940.* Princeton: Princeton University Press, 1983.

————. "Minería y agricultura en la sierra central: Formación y trayectoria de una clase dirigente regional, 1830–1910." In *Lanas y capitalismo en los Andes centrales,* ed. Florencia Mallon. Lima: Taller de Estudios Andinos, Universidad Nacional Agraria de La Molina, 1977.

————. *Peasant and Nation: The Making of Postcolonial Mexico and Peru.* Berkeley: University of California Press, 1995.

————, ed. "Lanas y capitalismo en los Andes centrales." Lima: Taller de Estudios Andinos, Universidad Nacional Agraria de La Molina, 1977 (mimeo).

Manrique, Manuel. "La colonización y la lucha por la tierra en el valle del Perené." In *Los Movimientos Campesinos en el Perú, 1879–1965,* Wilfredo Kapsoli, 267–300. Lima: Delva Editores, 1977.

Manrique, Nelson. "Los arrieros de la sierra central durante el siglo diecinueve." *Allpanchis* 18, 21 (Instituto de Pastoral Andina, Cusco, 1983): 27–46.

———. "Basadre y la guerra del Pacífico," In *Jorge Basadre: La política y la historia*, Noé Jave, 191–225. Lima: Lluvia Editores, 1981.

———. "El desarrollo del mercado interior en la sierra central, 1830–1910." Lima: Taller de Estudios Andinos, Universidad Agraria de La Molina, 1979 (mimeo).

———. *Las guerrillas indígenas en la guerra con Chile*. Lima: Centro de Investigación y Capacitación, 1981.

———. "La historiografía peruana sobre el siglo diecinueve." *Revista Andina* 9, 1 (Cusco, 1991): 241–59.

———. *Mercado interno y región: La sierra central, 1820–1930*. Lima: DESCO, 1987.

———. *Yawar Mayu: Sociedades terratenientes serranas, 1879–1910*. Lima: DESCO and Instituto Francés de Estudios Andinos, 1988.

Mantoux, Paul. *La révolution industrielle au dix-huitième siècle*. Paris: Editions Génin, 1959.

Martínez Alier, Joan. *Los Huacchilleros del Perú: Dos estudios de formaciones sociales agrarias*. Lima and Paris: Instituto de Estudios Peruanos and Ruedo Ibérico, 1973.

Martínez-Vergne, Teresita. *Capitalism in Colonial Puerto Rico: Central San Vicente in the Late Nineteenth Century*. Gainesville: University Press of Florida, 1992.

Marx, Karl. *Capital: A Critique of Political Economy*. 3 vols. New York: International Publishers, 1967.

Mathew, William M. "Anthony Gibbs and Sons, the Guano Trade, and the Peruvian Government, 1842–1861." In *Business Imperialism, 1840–1930: An Inquiry Based on British Experience in Latin America,* ed. D. C. M. Platt, 337–70. Oxford: Clarendon Press, 1977.

———. "The Imperialism of Free Trade: Peru, 1820–1870," in *Economic History Review* 21, 3 (London, 1968): 562–79.

———. *The House of Gibbs and the Peruvian Guano Monopoly*. London: Royal Historical Society, 1981.

———. *Anglo-Peruvian Commercial and Financial Relations, 1820–1865*. Ph. D. dissertation, London: University of London, 1964.

Matos Mar, José, ed. *Hacienda, comunidad, y campesinado en el Perú.* Lima: Instituto de Estudios Peruanos, 1976.

Matos Mar, José, and Fernando Fuenzalida. "Proceso de la sociedad rural." In *Hacienda, comunidad, y campesinado en el Perú,* ed. José Matos Mar, 15–50. Lima: Instituto de Estudios Peruanos, 1976.

Maude, H. E. *Slavers in Paradise: The Peruvian Slave Trade in Polynesia, 1862–1864.* Stanford: Stanford University Press, 1981.

McArver, Charles. "Mining and Diplomacy: United States Interests at Cerro de Pasco, 1876–1930." Ph.D. dissertation, University of North Carolina, 1977.

Mendels, Franklin. "Des industries rurales à la protoindustrialisation: Historique d'un changement de perspective." *Annales: Economies, sociétés, civilizations* 39, 5 (Paris, September–October 1984): 977–1008.

Mendiburu, Manuel de. *Diccionario histórico-biográfico del Perú.* Lima: Librería e Imprenta Gil, 1934.

Meuvret, Jean. *Etudes d'histoire économique.* Paris: Librairie Armand Colin, 1971.

Miller, Rory. "The Making of the Grace Contract: British Bondholders and the Peruvian Government, 1885–1890." *Journal of Latin American Studies* 8 (1976): 73–100.

———. "Railways and Economic Development in Central Peru, 1890–1930." In *Social and Economic Change in Modern Peru,* ed. Rory Miller et al. Liverpool: University of Liverpool, 1976.

———, ed. *Region and Class in Modern Peruvian History.* Liverpool: University of Liverpool, 1987.

Miller, Rory, Clifford T. Smith, and John Fisher, eds. *Social and Economic Change in Modern Peru.* Liverpool: University of Liverpool, 1976.

Miño Grijalva, Manuel. "Capital comercial y trabajo textil: Tendencias generales de la protoindustria colonial latinoamericana." *Hisla,* no. 9 (Revista Latinoamericana de Historia Económica y Social, Lima, 1987): 59–79.

Miró Quesada, Francisco, Franklin Pease, and David Sobrevilla, eds. *Historia, problema, y promesa: Homenaje a Jorge Basadre.* 2 vols. Lima: Pontificia Universidad Católica del Perú, 1978.

Mitre, Antonio. "Economic and Social Structure of Silver Mining in Nineteenth-Century Bolivia." Ph.D. dissertation, Columbia University, 1977.

———. *Los patriarcas de la plata: Estructura socioeconómica de la minería boliviana en el siglo diecinueve.* Lima: Instituto de Estudios Peruanos, 1981.

Montoya, Rodrigo. *Capitalismo y no capitalismo en el Perú: Un estudio histórico de su articulación en un eje regional.* Lima: Mosca Azul Editores, 1980.

Moreyra y Paz Soldán, Manuel. *La moneda colonial en el Perú: Capítulos de su historia.* Lima: Banco Central de Reserva del Perú, 1980.

Morin, Francoise, ed. *Indianité, ethnocide, indigenisme en Amérique latine.* Toulouse: Centre National de la Recherche Scientifique, 1982.

Morineau, Michel. *Incroyables gazettes et fabuleux métaux: Les rétours des trésors américaines d'après les gazettes hollandaises (seizième–dix-huitième siècles).* Paris and London: Editions de la Maison des Sciences de l'Homme and Cambridge University Press, 1985.

Morner, Magnus. *Notas sobre el comercio y los comerciantes del Cusco desde fines de la colonia hasta 1930.* Lima: Instituto de Estudios Peruanos, 1979.

Morse, Richard M. *El espejo de Próspero.* Mexico City: Siglo Veintiuno Editores, 1982.

———. "The Heritage of Latin America." In *The Founding of New Societies,* ed. Louis Hartz. New York: Mentor Books, 1964.

Murra, John V. *Formaciones económicas y políticas del mundo andino.* Lima: Instituto de Estudios Peruanos, 1975.

———. *La organización económica del estado inca.* Mexico City: Siglo Veintiuno Editores, 1978.

Novick, Peter. *That Noble Dream: The "Objectivity Question" and the American Historical Profession.* Cambridge: Cambridge University Press, 1988.

Núñez, Estuardo, ed. *El Perú visto por viajeros.* 2 vols. Lima: Ediciones Peisa, 1973.

Núñez, Lautaro. "L'évolution millénaire d'une vallée: Peuplement et ressources à Tarapacá." *Annales: Economies, sociétés, civilizations,* 33e année, n.s., 5–6 (Paris, 1978): 906–20.

Oficialía Mayor de Cultura. *Platería civil*. La Paz: Museos Municipales, 1992.

Orellana Muermann, Marcela, and Juan G. Muñoz Correa, eds. *Mundo minero: Chile, siglos diecinueve y veinte*. Santiago: Universidad de Santiago de Chile, 1991.

Orlove, Benjamin. "Alpaca, Sheep, and Men: The Wool Export Economy and Regional Society in Southern Peru." Ph.D. dissertation, University of California, Berkeley, 1974.

Pacheco, Mariano, Miguel Salcedo, and Toribio Yantas. "Pasco en la Colonia, siglos dieciséis, diecisiete, y dieciocho." In *Pasco colonial*. Cerro de Pasco: Universidad Nacional Daniel Alcides Carrión, 1980, 1–38 (mimeo).

Padilla Bendezú, Abraham. "Historia de la inmigración en el Perú." In *Juan de Arona y la inmigración en el Perú*, ed. Jorge Guillermo Llosa, 217–62. Lima: Academia Diplomática del Perú, 1971.

Palomeque, Silvia. "Loja en el mercado interno colonial." *Hisla*, no. 2 (Revista Latinoamericana de Historia Económica y Social, Lima, 1983): 33–45.

Paris, Robert. "El marxismo de Mariátegui." *Aportes* 17 (Revista de Estudios Latinoamericanos, Paris, July 1970).

Pease, Franklin. *Del Tawantinsuyu a la historia del Perú*. Lima: Instituto de Estudios Peruanos, 1978.

Pérez Arauco, César. "Anales del Cerro de Pasco: Referencias cronológicas de nuestra historia." *El Pueblo*, no. 18 (Revista Cultural de Difusión Popular, Cerro de Pasco, November 1978).

———. *Cerro de Pasco: Historia del pueblo mártir del Perú, siglos dieciséis, diecisiete, dieciocho, y diecinueve*. Cerro de Pasco: Edición de El Pueblo, 1980.

Piel, Jean. *Capitalisme agraire au Pérou: L'essor du neo-latifundisme dans le Pérou republicain*. Paris: Editions Anthropos, 1983.

Platt, D. C. M. "Dependency in Nineteenth-Century Latin America: An Historian Objects." *Latin American Research Review* 15, 1 (1980): 113–30.

———. "The Imperialism of Free Trade: Some Reservations." *Economic History Review* 21 (1968): 296–306.

————. *Latin America and British Trade, 1806–1914.* New York: Harper and Row, 1973.

————, ed. *Business Imperialism, 1840–1930: An Inquiry Based on British Experience in Latin America.* Oxford: Clarendon Press, 1977.

Platt, Tristan. "The Andean Experience of Bolivian Liberalism, 1825–1900: Roots of Rebellion in Nineteenth-Century Chayanta (Potosí)." In *Resistance, Rebellion and Consciousness in the Andean Peasant World, Eighteenth to Twentieth Centuries,* ed. Steve J. Stern, 280–323. Madison: University of Wisconsin Press, 1987.

————. "The Ayllus of Lipez in the Nineteenth Century." Paper presented at the forty-fourth Americanists International Congress, Manchester, September 1982.

Proctor, Robert. "Cerro de Pasco y la explotación minera." In *El Perú visto por viajeros,* ed. Estuardo Núñez, 2:24–35. 2 vols. Lima: Ediciones Peisa, 1973.

Quiroz, Alfonso W. *Banqueros en conflicto: Estructura financiera y economía peruana, 1884–1930.* Lima: Universidad del Pacífico, 1990.

————. *Domestic and Foreign Finance in Modern Peru, 1850–1950: Financing Visions of Development.* Pittsburgh: University of Pittsburgh Press, 1993.

Ramírez, Susan E. *Provincial Patriarchs: Land Tenure and the Economics of Power in Colonial Peru.* Albuquerque: University of New Mexico Press, 1986.

Ramírez Gastón, Javier, and Soledad Arispe, eds. *Apuntes sobre el proceso histórico de la moneda. Perú, 1820–1920.* Lima: Banco Central de Reserva del Perú, 1993.

Ramos, Carlos Augusto. *Toribio Pacheco: Jurista peruano del siglo diecinueve.* Lima: Pontificia Universidad Católica del Perú, 1993.

Ravines, Roger, ed. *Tecnología andina.* Lima: Instituto de Estudios Peruanos and Instituto de Investigación Tecnológica Industrial y de Normas Técnicas, 1978.

Reddy, William M. *The Rise of Market Culture: The Textile Trade and French Society, 1750–1900.* Cambridge and Paris: Cambridge University Press and Editions de la Maison des Sciences de l'Homme, 1984.

Regal, Alberto. *Historia de los ferrocarriles de Lima.* Lima: Universidad Nacional de Ingeniería, 1965.

Rénique, Gerardo. "El desarrollo de una empresa ganadera en los Andes centrales (1910–1960)." *Tierra y sociedad* 1, 1 (Revista del Archivo del Fuero Agrario, Lima, April 1978): 39–59.

———. "Sociedad ganadera del centro: Pastores y sindicalización en una hacienda alto-andina." Lima: Taller de Estudios Andinos, Universidad Nacional Agraria de La Molina, 1977 (mimeo).

Reyes Flores, Alejandro. *Contradicciones en el Perú colonial: Región central, 1650–1810.* Lima: Universidad Nacional Mayor de San Marcos, 1983.

———. "Estudios socio-económicos de los pueblos de Pasco, siglo dieciocho." In *Pasco colonial.* Cerro de Pasco: Universidad Nacional Daniel Alcides Carrión, 1980, 39–88 (mimeo).

Rodríguez Achung, Martha. "Interpretación de la historia político-sindical del proletariado siderúrgico, 1957–1972." Lima: Pontificia Universidad Católica del Perú, Taller de Estudios Urbano Industriales, 1980 (mimeo).

Rodríguez Pastor, Humberto. "Los trabajadores chinos culíes en el Perú: Artículos históricos." Lima: n.p., 1977 (mimeo).

Romano, Ruggiero. "American Feudalism." *Hispanic American Historical Review* 64, 1 (1984): 121–34.

———. "Fundamentos del funcionamiento del sistema económico colonial." In *El sistema colonial en la América española,* ed. Heraclio Bonilla, 239–80. Barcelona: Editorial Crítica, 1991.

———. *Les mécanismes de la conquête coloniale: Les conquistadores.* Paris: Flammarion, 1972.

———. "Sous-développement économique et sous-développement culturel, à propos d'André Gunder Frank." *Cahiers Vilfredo Pareto* 24 (Geneva, 1971): 271–79.

Ross, Dorothy. "Grand Narrative in American Historical Writing: From Romance to Uncertainty." *American Historical Review* 100, 3 (June 1995): 651–77.

Rostworowski, María. *Etnía y sociedad: Costa peruana prehispánica.* Lima: Instituto de Estudios Peruanos, 1977.

———. *Recursos naturales renovables y pesca, siglos dieciséis y diecisiete.* Lima: Instituto de Estudios Peruanos, 1981.

Salas de Coloma, Miriam. *De los obrajes de Canaria y Chincheros a las comunidades indígenas de Vilcashuamán, siglo dieciseis.* Lima: n.p., 1979.

Salvucci, Richard J. *Textiles and Capitalism in Mexico: An Economic History of the Obrajes, 1539–1840.* Princeton: Princeton University Press, 1987.

Samamé Boggio, Mario. *El Perú minero.* Lima: Instituto Geológico, Minero, y Metalúrgico, 1981.

Samuels, Warren J., ed. *The Economy as a System of Power.* New Brunswick, N.J.: Transaction Books, 1979.

Sánchez, Luis Alberto. *Historia de una industria peruana: Cervecería Backus y Johnston, S.A..* Lima: Editorial Científica, 1978.

Sánchez Albornoz, Nicolás. "La extracción de mulas de Jujuy al Perú: Fuentes, volumen, y negociantes." *Estudios de historia social,* no. 1 (Buenos Aires, 1965): 107–20.

———. "La saca de mulas de Salta al Perú, 1778–1808." *Anuario del Instituto de Investigaciones Históricas,* no. 8 (Rosario, Argentina, 1965): 261–312.

Sánchez Barba, Juan. "La via terrateniente y campesina en el desarrollo capitalista en la sierra central: El caso de Cerro de Pasco." In *Campesinado y capitalismo,* ed. Juvenal Casaverde et al., 147–234. Huancayo, Peru: Instituto de Estudios Andinos, 1979.

Saragoza, Alex M. *The Monterrey Elite and the Mexican State, 1880–1940.* Austin: University of Texas Press, 1988.

Shanin, Theodore, ed. *Peasants and Peasant Societies.* Baltimore: Penguin, 1971.

Smith, Clifford T. "Patterns of Urban and Regional Development in Peru on the Eve of the Pacific War." In *Region and Class in Modern Peruvian History,* ed. Rory Miller, 77–101. Liverpool: University of Liverpool, 1987.

Starr, Chester G. *A History of the Ancient World.* 4th ed. New York: Oxford University Press, 1991.

Stearns, Peter N. *Interpreting the Industrial Revolution.* Washington, D.C.: American Historical Association, 1991.

Stein, Barbara, and Stanley Stein. "D. C. M. Platt, the Anatomy of 'Autonomy.'" In *Latin American Research Review* 15, 1 (1980): 131–46.

Stern, Steve J. *Peru's Indian Peoples and the Challenge of Spanish Conquest. Huamanga to 1640.* 2d ed. Madison: University of Wisconsin Press, 1993.

———, ed. *Resistance, Rebellion, and Consciousness in the Andean Peasant World, Eighteenth to Twentieth Centuries.* Madison: University of Wisconsin Press, 1987.

Stewart, Watt. *Chinese Bondage in Peru: A History of the Chinese Coolie in Peru, 1849–1874.* Durham, N.C.: Duke University Press, 1951.

———. *Henry Meiggs, Yankee Pizarro.* Durham, N.C.: Duke University Press, 1946.

Sulmont, Denis. "Historia del movimiento obrero minero metalúrgico (hasta 1970)." *Tarea*, no. 2 (Revista de Cultura, Lima, October 1980).

Százdi, Adam. "Credit—Without Banking—in Early-Nineteenth-Century Puerto Rico." *Americas* 19, 2 (October 1962): 149–71.

Tamayo Herrera, José. *Historia del indigenismo cuzqueño, siglos dieciséis-veinte.* Lima: Instituto Nacional de Cultura, 1980.

Tandeter, Enrique. *Coacción y mercado: La minería de la plata en el Potosí colonial, 1692–1826.* Cusco: Centro de Estudios Regionales Andinos Bartolomé de las Casas, 1992.

———. *Coercion and Market: Silver Mining in Colonial Potosí, 1692–1826.* Albuquerque: University of New Mexico Press, 1993.

Tandeter, Enrique, Vilma Milletich, María Matilde Ollier, and Beatriz Ruibal. "El mercado de Potosí a fines del siglo dieciocho." In *La Participación Indígena en los Mercados Surandinos. Estrategias y Reproducción Social, siglos dieciséis al veinte,* ed. Olivia Harris et al., 379–424. La Paz: Centro de Estudios de la Realidad Social y Económica, 1988.

Tandeter, Enrique, and Nathan Wachtel. "Conjonctures inverses: Le mouvement des prix à Potosí pendant le dix-huitième siècle." *Annales: Economies, sociétés, civilizations* 38, 3 (Paris, May–June 1983): 549–613.

Tantalean Arbulú, Javier. *Política económico-financiera y la formación del estado, siglo diecinueve.* Lima: Centro de Estudios para el Desarrollo y la Participación, 1983.

Tarnawiecki, Donald. "Crisis y desnacionalización de la minería peruana: El caso de Cerro de Pasco, 1880–1901." Tesis de economía, Pontificia Universidad Católica del Perú, Lima, 1978.

Tauro del Pino, Alberto. *Diccionario enciclopédico del Perú.* 3 vols., app. Lima: Editorial Juan Mejía Baca, 1967.

Taylor, Lewis. "Earning a Living in Hualgayoc, 1870–1900." In *Region and Class in Modern Peruvian History,* ed. Rory Miller, 103–24. Liverpool: University of Liverpool, 1987.

———. "Main Trends in Agrarian Capitalist Development: Cajamarca, Peru, 1880–1976." Ph.D. dissertation, University of Liverpool, 1980.

Thompson, Slason. *A Short History of American Railways.* New York: D. Appleton, 1925.

Thorner, Daniel. "L'économie paysanne: Concept pour l'histoire économique." *Annales: Economies, sociétés, civilizations* 19, 3 (Paris, May–June, 1964), 417–32.

Thorp, Rosemary, and Geoffrey Bertram. *Peru 1890–1977: Growth and Policy in an Open Economy.* London: Macmillan, 1978.

Tord, Javier, and Carlos Lazo. *Hacienda, comercio, fiscalidad, y luchas sociales (Perú colonial).* Lima: Biblioteca Peruana de Historia, Economía, y Sociedad, 1981.

Tord, Luis Enrique. *El indio en los ensayistas peruanos, 1848–1948.* Lima: Editoriales Unidas, 1978.

Torero, Alfredo. *El Quechua y la historia social andina.* Lima: Universidad Ricardo Palma, 1974.

Trazegnies, Fernando de. *La idea del derecho en el Perú republicano del siglo diecinueve.* Lima: Pontificia Universidad Católica del Perú, 1980.

Urrutia, Jaime. "Comerciantes, arrieros, y viajeros huamanguinos, 1770–1870." Tesis de antropología, Universidad Nacional San Cristóbal de Huamanga, Ayacucho, 1982.

———. "De las rutas, ferias, y circuitos en Huamanga." *Allpanchis* 18, 21 (Instituto de Pastoral Andina, Cusco, 1983): 47–64.

Valderrama, Ricardo, and Carmen Escalante. "Arrieros, troperos, y llameros en Huancavelica." *Allpanchis* 18, 21 (Instituto de Pastoral Andina, Cusco, 1983): 65–88.

Vayssiere, Pierre. *Un siècle de capitalisme minier au Chili, 1830–1930.* Toulouse: Centre National de la Recherche Scientifique, 1980.

Vilar, Pierre. *La Catalogne dans l'Espagne moderne: Recherches sur les fondements économiques des structures nationales.* 3 vols. Paris: SEVPEN, 1962.

———. *Crecimiento y desarrollo: Economía e historia: Reflexiones sobre el caso español.* 3d ed. Barcelona: Editorial Ariel, 1976.

———. *Or et monnaie dans l'histoire.* Paris: Flammarion, 1974.

Villalobos, Sergio. *Comercio y contrabando en el Río de la Plata y Chile.* Buenos Aires: Editorial Universitaria, 1965.

Villanueva, Víctor. *Ejército peruano: Del caudillaje anárquico al militarismo reformista.* Lima: Librería–Editorial Juan Mejía Baca, 1973.

Volk, Steven S. "Crecimiento sin desarrollo: Los propietarios mineros chilenos y la caida de la minería en el siglo diecinueve." In *Minería americana colonial y del siglo diecinueve,* ed. Inés Herrera Canales and Rina Ortiz Peralta, 69–118. Mexico City: Instituto Nacional de Antropología e Historia, 1994.

Wachtel, Nathan. *Le rétour des ancêtres: Les Indiens urus de Bolivie, vingtième-seizième siècles: Essai d'histoire régressive.* Paris: Gallimard, 1990.

———. *La vision des vaincus: Les Indiens du Pérou devant la conquête espagnole, 1530–1570.* Paris: Gallimard, 1971.

Walker, Charles. "Montoneros, bandoleros, malhechores: Criminalidad y política en las primeras décadas republicanas." In *Bandoleros, abigeos, y montoneros: Criminalidad y violencia en el Perú, siglos dieciocho-veinte,* ed. Carlos Aguirre and Charles Walker, 105–36. Lima: Instituto de Apoyo Agrario, 1990.

Walker, David W. *Kinship, Business, and Politics: The Martínez del Río Family in Mexico, 1824–1867.* Austin: University of Texas Press, 1986.

Weber, Max. *The Protestant Ethic and the Spirit of Capitalism.* Trans. from the German by Talcott Parsons. New York: Scribner's, 1958. Original edition, Tübingen: J. C. B. Mohr, 1904–5.

Wilson, Fiona. "The Dynamics of Change in an Andean Region: The Province of Tarma, Peru, in the Nineteenth Century." Ph.D. dissertation, University of Liverpool, 1978.

———. "Propiedad e ideología: Estudio de una oligarquía en los Andes centrales (siglo diecinueve)." *Análisis* 8–9 (Cuadernos de Investigación, Lima, 1979): 36–54.

Wolf, Eric R. "Closed Corporate Peasant Communities in Mesoamerica and Central Java." *Southwestern Journal of Anthropology* 13 (1957): 1–18.

———. *Peasants*. Englewood Cliffs, N.J.: Prentice Hall, 1966.

———. "Types of Latin American Peasantry: A Preliminary Analysis." *American Anthropologist* 57 (June 1955): 452–71.

Wood, Gordon S. "A Century of Writing Early American History: Then and Now Compared; or, How Henry Adams Got It Wrong." *American Historical Review* 100, 3 (June 1995): 678–96.

Wu Brading, Celia. *Generales y diplomáticos: Gran Bretaña y el Perú, 1820–1840*. Lima: Pontificia Universidad Católica del Perú, 1993.

Yepes del Castillo, Ernesto. *Perú 1820–1920: Un siglo de desarrollo capitalista*. Lima: Instituto de Estudios Peruanos and Campodónico Ediciones, 1972.

Zapata, Gastón Antonio. "La crise de l'état national au Pérou pendant la guerre du Pacifique, 1879–1883." 2 vols. Thèse d'histoire, Ecole des Hautes Etudes en Sciences Sociales, Paris.

Index